S0-ATC-037

Forensic Engineering

Lethbridge Community College Library

Forensic Engineering

Editor

Kenneth L. Carper, Architect, M.ASCE
Professor, School of Architecture
Washington State University
Pullman, Washington

CRC Press

Boca Raton Boston London New York Washington, D.C.

Library of Congress Cataloging-in-Publication Data

Forensic engineering / editor, Kenneth L. Carper.
        p. cm.
    Includes index.
    ISBN 0-8493-7483-9
    1. Forensic engineering. 2. Structural failures—Investigation.
  I. Carper, Kenneth L.
  TA656.F66 1998
  620—dc19   88-21254
                                                               CIP

This book contains information obtained from authentic and highly regarded sources. Reprinted material is quoted with permission, and sources are indicated. A wide variety of references are listed. Reasonable efforts have been made to publish reliable data and information, but the author and the publisher cannot assume responsibility for the validity of all materials or for the consequences of their use.

Neither this book nor any part may be reproduced or transmitted in any form or by any means, electronic or mechanical, including photocopying, microfilming, and recording, or by any information storage or retrieval system, without prior permission in writing from the publisher.

The consent of CRC Press LLC does not extend to copying for general distribution, for promotion, for creating new works, or for resale. Specific permission must be obtained in writing from CRC Press LLC for such copying.

Direct all inquiries to CRC Press LLC, 2000 Corporate Blvd., N.W., Boca Raton, Florida 33431.

Trademark Notice: Product or corporate names may be trademarks or registered trademarks, and are used only for identification and explanation, without intent to infringe.

© 1989 by Elsevier Science Publishing Co., Inc.
© 1998 by CRC Press LLC

No claim to original U.S. Government works
International Standard Book Number 0-8493-7483-9
Library of Congress Card Number 88-21254
Printed in the United States of America   3 4 5 6 7 8 9 0
Printed on acid-free paper

Contents

14. Conclusion   347
KENNETH L. CARPER

Preface

This book is an overview of the activities of forensic experts in the engineering professions. General chapters cover aspects of forensic activity that are common to all disciplines. Specific chapters, contributed by experts in each specialized field, detail unique aspects of forensic engineering and accident reconstruction in the various specialized disciplines.

Contributors have been carefully selected to represent a wide variety of significant experience in forensic practice. Biographical information regarding contributors is given in the section immediately following. Each chapter is a brief informative overview and is intended to be a reflection primarily of the contributor's own experiences. Viewpoints expressed in a particular chapter are the author's own, and do not necessarily represent the viewpoints of the publisher, the editor, or other contributors.

Contributors have included the following information as it relates to their specific disciplines:

Typical clients, and scope and purpose of investigations

Techniques, procedures, and tools used in investigation and analysis

Interface with forensic specialists from other disciplines

Impact of forensic activity on improved practices, products, or planning to reduce the frequency and severity of failures

Case study examples from the contributors' experiences

Reference lists for further reading

Some contributors have included reference to basic theories used in a typical analysis; others have focused on the organizational aspects of an investigation, or on the institutional framework for the practice of forensic engineering.

This book is the first broad overview of forensic activity and accident investigation in engineering. It is expected to appeal to a varied audience which will include forensic experts practicing in all engineering disciplines, design and construction professionals, attorneys, product manufacturers, insurance professionals, and engineering students.

Contributors

Glenn R. Bell is a Senior Associate of Simpson Gumpertz & Heger Inc. of Arlington, Massachusetts, and San Francisco, California, where he is involved principally with investigation of structural distress and failure, and the design of unusual and conventional structures. He holds a bachelor's degree in civil engineering from Tufts University, where he has also taught structural engineering, and a master's degree in structural engineering and structural mechanics from the University of California at Berkeley. His extensive experience in failure investigation includes acting as principal investigator in his firm's investigation into the walkways collapse at the Hyatt Regency in Kansas City, Missouri. He has had direct responsibility for major structural design projects, including "Spaceship Earth," the geodesic sphere at Walt Disney World Epcot Center. Bell is a registered professional engineer and is active in several professional societies. He is vice chairman of the ASCE Technical Council on Forensic Engineering, a founding member and past chairman of the ASCE Committee on Dissemination of Failure Information, and a member of the editorial board of the ASCE *Journal of Performance of Constructed Facilities*. He has lectured and written extensively on the subject of structural design and the investigation of structural failure.

Kenneth L. Carper is a registered architect with a professional bachelor of architecture degree as well as a master's degree in structural engineering. He is a member of the American Society for Engineering Education and the American Society of Civil Engineers. Carper is a founding member and past chairman of the ASCE Committee on Dissemination of Failure Information. He currently serves as secretary on the Executive Committee of the ASCE Technical Council on Forensic Engineering (TCFE) and as chairman of the TCFE Publications Committee. He is the founding editor of the ASCE/AEPIC/NSPE *Journal of Performance of Constructed Facilities*. Carper is an associate professor in the School of Architecture at Washington State University, where he has the responsibility for all architectural structures courses. He has received several teaching awards, including the annual award for Outstanding Professor in the College of Engineering and Architecture. Carper has written several papers on forensic engineering and has coordinated the publication of a book on forensic civil engineering. He has received a national award from the ASCE for one of his publications. Carper has lectured extensively about structural failures to students, faculty, and professional groups of architects, engineers, and building officials.

Ron Hendry is a registered professional engineer in Texas. He has a degree in mechanical engineering from Texas A&M University. He has 17 years of product design experience with several companies serving diverse industries. Since 1980, Hendry has been a practicing forensic engineer, specializing in accident reconstruction of machinery and vehicles, and product failure analysis. He is a member of the National Academy of Forensic Engineers. His other organizational memberships include the NSPE, TSPE, and SAE.

Joel T. Hicks is president of General Technology, a firm that provides consulting mechanical engineering services in the fields of engineering management, product and process development, product safety and reliability review, and forensic investigations. He is a partner in a firm that has developed computer-aided solutions for accident reconstruction and related technical investigations. Hicks is a registered professional engineer and is active in several professional societies, including the ASME, NSPE, and SAE. He was a charter member and director of the National Academy of Forensic Engineers.

William G. Hyzer is one of the world's leading authorities in photographic instrumentation—the technology of applying photography as an instrument of scientific investigation. He is a specialist in photographic measurement, including image analysis, image enhancement, and photographic interpretation for scientific and forensic applications. Hyzer holds degrees from the Universities of Minnesota and Wisconsin. He is the author of two scientific books and over 500 published technical articles. He is a registered professional engineer and has been in private practice since 1953 as a consulting physicist and engineer in Janesville, Wisconsin. He is also the author of a regular column in the monthly publication *Photomethods* and serves on the editorial board of *Industrial Research & Development* magazine. Hyzer is a fellow of both the Society of Photo-

Optical Instrumentation Engineers and the Society of Motion Picture and Television Engineers, and was national vice president of the SMPTE. He has received several national and international awards as recognition for significant contributions in engineering and photography.

Rudolf Kapustin is president and principal consultant of International Aviation Safety Consultants (INTASCO), a firm that provides technical consultations in all matters of air safety, accident investigations, investigative procedures, and emergency preparedness for aviation insurance underwriters, law firms, and aerospace industry clients. Kapustin served the Civil Aeronautics Board and the National Transportation Safety Board for 24 years, during which time he had overall responsibility for the investigation of catastrophic air carrier accidents. He is a graduate of the Academy of Aeronautics in New York. Kapustin was employed in aircraft line maintenance and field service engineering positions with a major airline and an aircraft engine manufacturer for 16 years before entering government service. He also lectures at major universities in courses on safety management and aviation accident investigation. Kapustin is a charter member and international secretary of the International Society of Air Safety Investigators.

Lindley Manning is a registered mechanical engineer in Nevada and Ohio. His degrees are from the Universities of Cincinnati and Nevada. He is a retired associate professor of mechanical engineering at the University of Nevada at Reno. Manning is a practicing forensic engineer, specializing in product liability litigation. He is a charter member of the National Academy of Forensic Engineers and serves on its national board of directors, currently as vice president. He is also an active member of numerous other professional societies, including the NSPE, ASME, and SAE.

M. D. Morris is a communications consultant to federal and local governments, major industry, and professional offices. He has taught over 550 writing courses to over 13,000 persons in the past 22 years. Morris is the editor of the Wiley Series of Practical Construction Guides (42 volumes) and is the author of over 500 published works including articles, technical papers, monographs, and four books. He is a registered professional engineer and a fellow of the American Society of Civil Engineers. Morris is the founder and past editorial chairman of the ASCE *Journal of the Construction Division* and the past chairman of the executive committee of the Construction Division of ASCE. He is also a senior member of the Society for Technical Communication and an active member of the Overseas Press Club of America.

Paul E. Pritzker is a 1948 graduate of Wentworth Institute of Technology. He received an honorary doctor of engineering technology degree in 1985 when he delivered the commencement address on "Engineering Ethics." Pritzker continues to serve as a member of the Wentworth Corporation. Pritzker is the president and senior engineer of George Slack & Pritzker, a technical organization with expertise in product safety analysis and fire investigation. He is registered as a professional engineer in the dual disciplines of electrical and safety engineering in California, Connecticut, Florida, Massachusetts, and Wisconsin. Pritzker serves as a director of the National Association of Fire Investigators and is a member of the Fire/Explosion Certification Board. He served as president of the National Society of Professional Engineers. He was a founder and the second president of the National Academy of Forensic Engineers. Pritzker is a member of numerous other professional societies, including the IEEE, ASSE, SFPE, and NFPA. He is an international lecturer and a prolific writer.

Fred H. Taylor is principal engineer and president of Taylor Systems Engineering, Inc., a facility and mechanical engineering firm located in Fair Oaks, California. His firm provides engineering services in forensic and systems analysis, energy conservation, mechanical systems design, and facility engineering in heating, ventilating, refrigerating, air conditioning, and related fields. Taylor is a registered professional engineer in mechanical and control systems engineering. He has been a member of the American Society of Heating, Refrigerating and Air Conditioning Engineers (ASHRAE) since 1953, serving as the Sacramento Valley Chapter president in 1965. He received a 1985 National ASHRAE Energy Award for the energy retrofit of the Graduate School of Business building at Stanford University in Palo Alto, California. He has written technical journal articles on energy conservation and has lectured on forensic and energy conservation topics. He has more than 35 years experience in the HVAC and mechanical engineering fields with extensive experience in HVAC system analysis and environmental systems failure investigations.

Joseph S. Ward maintains a consulting geotechnical engineering practice, specializing in forensic engineering, in Montclair, New Jersey. He is a registered professional engineer in 20 jurisdictions and is a licensed land surveyor in two states. Ward holds a B.C.E. degree from Manhattan College and a master's degree from Rutgers University. He has also undertaken graduate studies at Columbia and Purdue Universities. He is a fellow and past national president of the American Society of Civil Engineers, a fellow and past national director of the American Consulting Engineers Council and is currently the president of the National Academy of Forensic Engineers. He is a past chairman of the executive committee of the ASCE Technical Council on Forensic Engineering, and serves on many other committees of the ASCE. Ward has practiced as a geotechnical consultant for over 38 years and is the former president/CEO of Converse Consultants, a nationwide geotechnical engineering firm based in Pasadena, California. His forensic work has involved the investigations of hundreds of underground construction failures. He has been instrumental in resolving numerous conflicts during construction.

1. What Is Forensic Engineering?

KENNETH L. CARPER

1.1 INTRODUCTION

1.1.1 Definitions

The forensic engineer is a professional engineer who deals with the engineering aspects of legal problems. Activities associated with forensic engineering include determination of the physical or technical causes of accidents or failures, preparation of reports, and presentation of testimony or advisory opinions that assist in resolution of related disputes. The forensic engineer may also be asked to render an opinion regarding responsibility for the accident or failure.

Milton F. Lunch, former General Counsel to the National Society of Professional Engineers (NSPE), has provided the following comprehensive definition of forensic engineering:

> Forensic Engineering is the application of the art and science of engineering in the jurisprudence system, requiring the services of legally qualified professional engineers. Forensic engineering may include investigation of the physical causes of accidents and other sources of claims and litigation, preparation of engineering reports, testimony at hearings and trials in administrative or judicial proceedings, and the rendition of advisory opinions to assist the resolution of disputes affecting life or property.

A condensed definition is given by Marvin M. Specter, founding President of the National Academy of Forensic Engineers (NAFE):

> Forensic Engineering is the art and science of professional practice of those qualified to serve as engineering experts in matters before courts of law or in arbitration proceedings. [Specter 1987]

1.1.2   Accident Reconstruction

Failures and accidents involving injury, loss of life, or property damage nearly always generate controversy. Hence, the investigation of such events is usually associated with litigation or the threat of litigation.

Accident investigation and reconstruction, however, need not always be directly related to litigation. Sometimes the principal purpose of accident reconstruction is to determine causation so that the accident will not be repeated. For example, Chapter 7 discusses in detail the investigation of airline accidents by the National Transportation Safety Board (NTSB). These investigations do not involve traditional litigation. In fact, the work product of an investigation by the NTSB is not admissible in legal proceedings arising from the accident. Such use of NTSB reports is forbidden by statute. The overriding concern is for a thorough and timely investigation. Recommendations are quickly disseminated to involved parties with the goal of reducing the potential for repetition of the accident (Figure 1.1). Structural failure investigations conducted by the National Bureau of Standards (NBS) have a similar purpose. NBS investigations focus on technical causation and do not attempt to assign responsibility among the involved parties. Figures 1.2 and 1.3 show test apparatus and

Figure 1.1   Investigators examining the wreckage of a cargo plane after it crashed on Mt. Rainier in Washington State. (Courtesy NTSB.)

Figure 1.2 Test apparatus at National Bureau of Standards Center for Building Technology, Gaithersburg, Maryland. This test supported the investigation of the Kansas City Hyatt Hotel walkway failure in 1981.

analysis of a test result during the NBS investigation of the collapse of the Kansas City Hyatt hotel walkways in 1981 (USDC/NBS 1982).

The foregoing "accident reconstruction" activities do not fit within the given definitions of forensic engineering, because they exist separate from the litigation arena. These activities are included in this book, however, because forensic engineers are often involved in these activities and ultimately include the reports developed by such activities in their own investigations.

Also, it is important to note that although the typical forensic investigation is conducted under the threat of litigation, very few cases actually go to the courtroom. Each contributor to this book, no matter what discipline, has mentioned the fact that most cases are resolved prior to

Figure 1.3 Analysis of results from the tests conducted using the apparatus shown in Figure 1.2. The analyst is studying the failure of the box–beam connection that supported the pedestrian walkways.

litigation, often as the result of convincing work done by competent forensic engineers. Forensic engineers are at the forefront in developing alternative dispute resolution techniques. Often they serve as consultants in arbitration, mediation, minitrials, and other procedures. These activities are discussed in Chapter 13.

Thus, although litigation overshadows forensic work, and the threat of litigation influences the care and thoroughness of each investigation, the forensic engineer may actually spend very little time testifying in the courtroom setting.

1.1.3   Typical Clients and Projects

Typical clients include all parties affected by the accident or failure. Often the forensic engineer works for attorneys representing plaintiffs or defendants, who may be individual parties, corporations, or governmental agencies.

In some cases, the forensic engineer works independently, using consultants as required. In other cases, the investigation is conducted by a team. For example, a major NTSB airline accident investigation may involve more than 100 individuals. In such investigations the forensic engineer may act as a coordinator, synthesizing the work of many specialists.

Sometimes the accident under investigation is in the public spotlight, for example, a major industrial accident or a structural collapse involving loss of life. In these cases, the forensic engineer may be questioned by reporters from the news media while the investigation is underway. Usually, however, forensic work relates to costly performance problems that are less spectacular, such as leaking roofs and facades or minor vehicle accidents.

1.1.4   Influence on Improved Practices

Frequently in the course of a forensic investigation, product or procedural deficiencies are uncovered that extend beyond the specific case. One of the more rewarding aspects of forensic engineering is the opportunity to make recommendations on the basis of such an investigation. Forensic engineers have a role in disseminating information to design professionals to improve procedures and products so that failures or accidents are not repeated. This aspect of the profession may be compared with the role of the forensic medical expert, or pathologist, in collecting and disseminating information that affects public health and safety. A forensic medical investigation, although related to litigation, may result in recommendations that effect improvements in medical practice. In the same way, forensic engineers can have an impact on improved engineering design practices.

1.2   QUALIFICATIONS OF THE FORENSIC ENGINEER

The specific technical skills required of the forensic engineer vary with the discipline. Some general personal and professional characteristics, however, are prerequisites for successful forensic practice.

1.2.1   Technical Competency

First, the forensic engineer must have demonstrated competency in his or her specialized engineering discipline. Competency is the result of education and experience. A Professional Engineering license is desirable to qualify for courtroom testimony as a forensic engineer. The forensic engineer who has an extensive professional background, with many years of successful engineering practice, is likely to be more effective in courtroom testimony. Engineers with less experience may be more comfortable and effective as members of the investigative support team. Active membership in appropriate professional societies also contributes to the credibility of an expert witness. Significant professional society activity generally indicates that the forensic engineer enjoys the respect of peers in the profession.

1.2.2 Knowledge of Legal Procedures

Along with technical competency, the forensic engineer must have a working knowledge of legal procedures and the related vocabulary. The vocabulary used in litigation is quite specific. The forensic engineer who is not familiar with this language can do irreparable damage to an otherwise sound presentation. Likewise, written communications or reports, if prepared or released at an inappropriate time, can inadvertently affect litigation.

1.2.3 Detective Skills

The forensic engineer, acting as investigator, must possess certain detective skills. Diligence must be exercised in collecting pertinent facts from the field and from documents (ASCE/RCPS 1986). The quantity and reliability of available data vary widely with each case and with each discipline. For example, a greater quantity of detailed and reliable evidence is available for a major airline accident than for an automobile accident investigation. In the airline investigation, there are copious records, including voice recordings and flight data recordings. In the automobile accident investigation, the investigator has limited access to evidence, especially if the investigation occurs several years after the accident. The evidence may be limited to police reports, photographs, and written records of witness statements.

Much of the investigation involves interpretation of the data collected. During the analysis, the investigator must separate contributing factors from irrelevant items. Data will always be incomplete and, in some cases, redundant. Redundant evidence is always helpful as it allows the forensic engineer to cross-check conclusions. Sometimes conflicts exist in redundant data that must be resolved.

An important aspect of data collection is respect for protection of evidence. Physical evidence must be preserved at all cost. Damage to evidence, even minor alteration, can profoundly influence the outcome of the investigation and the outcome of related litigation.

Timing is critical to the collection of reliable data. The forensic engineer is often required to move quickly to an accident site. Evidence is sometimes destroyed immediately after a fire, industrial accident, or major structural collapse during the rescue and cleanup operations (Figure 1.4). Immediate response to such events requires a flexible and resilient personality.

Part of the detective work is research into design standards and material properties in existence at the time the failed product or project was created. In this sense, the forensic engineer acts as historian. The failure must be analyzed from the perspective of technical knowledge and professional procedures in a context that might have been much different from present practice or the investigator's own design standards.

Figure 1.4 L'Ambiance Plaza, Bridgeport, Connecticut. This project collapsed during construction in April 1987, killing 28 construction workers. Useful evidence is often destroyed during the rescue efforts after such an accident (USDC/NBS 1987, Scribner and Culver 1988).

The importance of detective skills, as opposed to design skills, cannot be overemphasized. A good design professional is not necessarily a good forensic expert. The designer is more familiar with codes and standards intended to prevent failure, and with the process of generating alternative concepts that respond to given constraints. The design-oriented engineer may approach the investigation of a failure by suggesting an alternative design solution. The forensic engineer, however, approaches the investigation of a failure from the perspective of physical causation and the given object, as designed and constructed. These two perspectives are much different and require different skills and attitudes.

1.2.4   Oral and Written Communication Skills

The forensic engineer must be an effective communicator, both in oral and in written presentations. During the investigation, the news media may interview the forensic engineer. Contacts with the media should be viewed as opportunities to restore public confidence and to show professional concern. This can be accomplished without making premature specific statements, if the forensic engineer speaks carefully and articulately.

Oral communication skills are also a prerequisite to effective testimony in the courtroom or in public hearings. The forensic engineer may perform as an educator, explaining complex technical issues in language that is understandable to laypersons with no technical background. The ability to use simple examples and clear language to illustrate complex phenomena is essential, and improves the credibility of the witness. Chapter 12 discusses the importance of oral communication skills in testimony.

A written report is usually produced after a forensic investigation. The report becomes the tangible record of the investigation. The quality of the report reflects on the competency of the investigator. Reports usually detail the conclusions of the investigation and include references to all relevant supporting data. Preparation of a successful report is the subject of Chapter 11.

1.2.5 Other Skills

Other skills are also helpful in forensic engineering. Familiarity with the fields of psychology and sociology is important. Photographic skills are very useful (Chapter 10). Often the forensic engineer retains consultants, such as imaging experts or human performance specialists, to provide specialized skills. Recognition of the need for particular specialized consultants is an important characteristic of the competent investigator.

1.2.6 Personality Characteristics

In addition to the foregoing skills, the effective forensic engineer or accident reconstructionist exhibits certain personality traits.

The most important characteristic for competent forensic engineering is high ethical standards. Ethical and professional principles are put to the test more often, and to a greater degree, than in any other engineering endeavor. The forensic engineer is in a position to influence adversely the professional and personal reputations of all involved parties. This position of influence must not be taken lightly. The forensic engineer must be able to maintain objectivity and impartiality in seeking truth, in the face of constant pressures to take an emotional or advocacy position. This important aspect of forensic practice is discussed further in the next section.

Flexibility is also a desirable personality characteristic. As stated earlier, the forensic engineer must be willing to travel, at a moment's notice, to the site of a failure or an accident. It is difficult to predict the scope or direction of an investigation at the outset. Adjustment to new demands and directions requires flexibility. Objectivity implies the ability to discard preconceived notions when the facts do not support preliminary hypotheses. Forensic practice is not a profession for stubborn "know-it-alls."

On the other hand, the effective forensic engineer must maintain confidence in his or her competence during legal proceedings. Cross-examination of expert witnesses often involves overt challenges to their personal integrity and professional competence. These challenges are unfamiliar to most engineers, and place a great deal of stress on those not prepared for the situation (Bleyl 1985).

Finally, the forensic engineer should have the ability to work effectively with others. Sometimes the forensic engineer is the coordinator of a team of investigators. In other cases, the forensic engineer may be a member of the team of support specialists. During the data collection and testing phase of an investigation, the forensic engineer may cooperate with other investigators representing diverse interests. Such cooperative ventures are extremely effective in reducing the cost and time associated with an investigation (Sowers 1986). Interpersonal skills are essential to maintaining positive working relationships under such conditions.

1.3 ETHICS AND PROFESSIONAL RESPONSIBILITIES

Litigation surrounding failures and accidents has reached an unprecedented level of activity in the past two decades. With this increased litigation has come an increased demand for engineering professionals to serve as expert witnesses. Many individuals have been attracted to the field of forensic engineering by the financial rewards and by the apparent limited exposure to liability, when compared with the increasing liability exposure facing the design engineer.

Unfortunately, some individuals who lack the skills and attitudes discussed in the previous section have entered this field. As a result of disreputable practices by unqualified individuals, the ethical principles of forensic engineers are often called into question. Several professional societies are working to maintain guidelines for ethical practices in forensic engineering. In fact, NAFE was established, in part, to address the need for ethical standards in forensic practice (Section 1.4).

NSPE has published *Guidelines for the Professional Engineer as a Forensic Engineer.* The Association of Soil and Foundation Engineers (ASFE) has prepared a document entitled *Recommended Practices for Design Professionals Engaged as Experts in the Resolution of Construction Industry Disputes"* (ICED 1988). This document, endorsed by a number of professional societies, is reproduced in the Appendix (Bachner 1988).

Ethical conflicts in forensic practice arise from the fact that forensic engineers are generally retained by parties to the dispute. Attorneys are retained to be advocates for their clients. Forensic engineers, however, must remain impartial "seekers of the truth," even when that truth is not in the client's interest. The forensic engineer is required, by ethical principles, to maintain objectivity. But the engineer's future income and reputation as a valuable expert witness depend on client satisfaction. Other potential conflicts of interest involve cases where the forensic engineer's own professional peers are questioned regarding technical competency.

Not all attorneys pressure forensic consultants to act in an advocacy role. Many attorneys actively seek experts with dissenting viewpoints (Worrall 1984). Adverse information is useful to the client and assists the decision-making process, resulting in settlement of the dispute. The impartial, objective forensic consultant serves the client well by pointing out weaknesses in the client's position. Such an approach is valued in the long run, and helps to establish a reputation for integrity and credibility.

Dissenting viewpoints among competent forensic engineers are common. Disagreement does not necessarily imply dishonesty or incompetence on the part of one of the witnesses. Often there is no single truth. Failures and accidents often result from a number of complex, interrelated factors. An honest expression of diversity of opinion, through the introduction of testimony by multiple experts, is a healthy approach to seeking truth. It is important to give consideration to all contributing factors.

Sometimes a controversial judicial decision is difficult to reconcile, given the facts presented in the case. This is frustrating to the dedicated forensic engineer who has expended considerable effort to support factual testimony. When reviewing a controversial decision, it is important to remember that the underlying purpose of litigation is to settle arguments, not to establish ultimate truth.

Ethical practices require the forensic engineer to be thorough, cross-checking all conclusions. Questions beyond those asked by the client should be addressed. All calculations should be studied for reasonable variances in data, and all ranges of uncertainty should be expressed.

Special care should be excercised in cases involving malpractice charges. The forensic engineer has a responsibility to protect the professional reputations of all parties until the investigation is complete. Unclear or premature statements to the media can irreparably damage the professional reputations of innocent parties. As stated earlier, where competency of the design professional is in question, it is the investigator's responsibility to determine the standard of practice at the time of design. Current standards and the forensic engineer's own design preferences are irrelevant to the question of competency.

All of these issues clearly test the professional integrity of the forensic consultant. The forensic engineer who performs these tasks admirably has a unique opportunity to represent the highest ideals of the engineering professions before the general public.

1.4 RESOURCES AND PROFESSIONAL ORGANIZATIONS

1.4.1 Professional Organizations

Several professional engineering and scientific societies have been active in supporting the interests of forensic engineers working in specialized disciplines. Some of the more active societies include the following:

Association of Soil and Foundation Engineers (ASFE): The Association of Engineering Firms Practicing in the Geosciences

American Society of Civil Engineers: Technical Council on Forensic Engineering (ASCE/TCFE)

International Civil Aviation Organization (ICAO)

International Society of Air Safety Investigators (ISASI)

National Academy of Sciences (NAS)

National Society of Professional Engineers (NSPE)

Society of Automotive Engineers (SAE)

In addition to the specialized organizations listed, there is one professional organization dedicated to the interests of forensic engineers from all disciplines. The National Academy of Forensic Engineers, established in 1982, maintains strict guidelines for membership and, among other activities, publishes a semiannual journal. NAFE is an outgrowth of discussions by NSPE. NAFE exists as an independent professional society, but remains a Chartered Affinity Group of NSPE (Specter 1987).

1.4.2 Journals and Regular Publications

Several organizations produce regular publications that contain valuable information for forensic engineers. These include the following:

ASFE Case Study Publications
(ASFE: The Association of Engineering Firms Practicing in the Geosciences)

Forensic Engineering
(Pergamon Press)

Forensic Science International
(Elsevier Science Publishers, Ireland)

Journal of Forensic Sciences
(American Academy of Forensic Sciences)

Journal of the National Academy of Forensic Engineers
(National Academy of Forensic Engineers)

Journal of Performance of Constructed Facilities
(American Society of Civil Engineers, in cooperation with the National Society of Professional Engineers and the Architecture and Engineering Performance Information Center)
SAE Technical Papers
(Society of Automotive Engineers)

1.5 OPPORTUNITIES IN FORENSIC ENGINEERING PRACTICE

Professional engineers who possess the qualifications and the desire to enter the field of forensic engineering are accepting a unique challenge. The financial rewards are attractive and the work is interesting, but the demands are great. Forensic engineering requires a commitment to ethical principles that are not often tested in traditional engineering practice. The competency and professional integrity of the forensic consultant are often challenged in adversarial situations. Forensic practice is unpredictable, requiring flexibility as a personality characteristic.

The principal contribution of the forensic engineer is the introduction of reason into the dispute settlement arena. In this profession, the practitioner can provide valuable assistance to the litigation process by reducing the confusion that usually accompanies a complex failure or accident. Forensic consultants can contribute to the development and implementation of alternate dispute resolution techniques, assisting in the settlement of disputes outside traditional litigation procedures (see Chapters 12 and 13).

The forensic engineer also has an opportunity to represent the professions of engineering before the general public. In the performance of an investigation, the forensic consultant can help to restore public confidence. Finally, the forensic engineer sometimes has the opportunity to make recommendations for improved practices, thereby leading to mitigation of the frequency and severity of failures and accidents.

References

American Society of Civil Engineers, Research Council on Performance of Structures (ASCE/RCPS). 1986. *Guide to Investigation of Structural Failures*, rev. ed. New York: ASCE.

Association of Soil and Foundation Engineers (ASFE). 1985. *Expert: A Guide to Forensic Engineering and Service as an Expert Witness.* Silver Springs, MD: ASFE.

Bachner, J. P. 1988. Facing Down the Hired Gun. *Journal of Performance of Constructed Facilities* (ASCE) 2, No. 4 (November).

Bleyl, R. L. 1985. The Forensic Engineer on the Witness Stand under Cross-Examination Attack. *Journal of the National Academy of Forensic Engineers* (NAFE) 2, No. 2 (December).

Carper, K. L. ed. 1986. *Forensic Engineering: Learning from Failures.* New York: ASCE.

Compressed Air. 1986. The forensic engineer. *Compressed Air* (Ingersoll-Rand Co., Washington, NJ) 91, No. 12 (December).

Interprofessional Council on Environmental Design (ICED). 1988. Recommended practices for design professionals engaged as experts in the resolution of construction industry disputes.

LePatner, B. B., and S. M. Johnson. 1982. *Structural and Foundation Failures: A Casebook for Architects, Engineers and Lawyers.* New York: McGraw–Hill.

Scribner, C. F., and C. G. Culver. 1988. Investigation of the Collapse of L'Ambiance Plaza. *Journal of Performance of Constructed Facilities* (ASCE) 2, No. 2 (May).

Sowers, G. F. 1986. Failure investigation for forensic engineering. In *Forensic Engineering: Learning from Failures.* New York: ASCE.

Specter, M. M. 1987. National Academy of Forensic Engineers. *Journal of Performance of Constructed Facilities* (ASCE) 1, No. 3 (August).

U.S. Department of Commerce, National Bureau of Standards (USDC/NBS). 1982. *Investigation of the Kansas City Hyatt Regency Walkways Collapse,* NBS Building Science Series 143. Washington, DC: U.S. Government Printing Office, May.

———— 1987. *Investigation of L'Ambiance Plaza Building Collapse in Bridgeport, Connecticut,* NBSIR 87-3640. Washington, DC: U.S. Government Printing Office, September.

Ward, J. S. 1986. What is a forensic engineer? In *Forensic Engineering: Learning from Failures.* New York: ASCE.

Worrall, D. G. 1984. Engineer experts—The attorney's viewpoint. *Journal of the National Academy of Forensic Engineers* (NAFE) 1, No. 2 (October).

2. Learning from Failures

KENNETH L. CARPER

2.1 INTRODUCTION

Much of the knowledge used to design, construct, manufacture, and operate engineered facilities and products has been obtained through learning from failures. Interdisciplinary communication about the causes of failures and accidents often results in improved design practices. This chapter briefly reviews the history of failure analysis as it relates to the evolution of engineering design. A definition of failure is given, and the root causes of failure are discussed.

The forensic engineer can make a significant contribution to the process of learning from failures by disseminating information to the design professions. Again, forensic engineering can be compared with the field of medical pathology, which has played a principal role in the development of medical science.

Professional and trade journals tend to emphasize successful projects and products, and rightly so. Modern engineering methods, construction procedures, and manufacturing processes result in many more success stories than failures. This is due, in large part, to the many dedicated professionals working to improve quality control in the design and production of engineered facilities and products.

The responsibility remains, however, for professionals to communicate about failures and accidents as well. The tradition of learning lessons from experience, including failure, continues to be important to the advancement of the engineering professions.

2.2   HISTORICAL CONTEXT: TRIAL AND ERROR TRADITION

The concept of learning from failures is fundamental to the practice of engineering in all fields. Design codes, standards of practice, and construction and manufacturing procedures have all evolved traditionally through a process of trial-and-error. For example, the development of engineering theory for the construction industry is relatively recent, when seen in the context of the history of construction.

In the past, builders based their designs on observations of performance of earlier construction. Failures led to a new understanding of structural behavior and to a corresponding improvement in design.

Robert Mark, in his research on cathedral design and construction during the Gothic period, has found evidence that a communication network existed among the various cathedral builders in Europe. This network allowed for dissemination of information about problems in performance so that corrective measures could be made on the various sites simultaneously (Mark 1982). This exchange of information was deemed essential in a period when builders were extending the frontiers of knowledge and experiencing behavior unpredicted by past efforts (Figure 2.1). Henry Petroski has noted that no one wants to learn by mistakes, but we cannot learn enough from successes to go beyond the state-of-the art (Petroski 1985).

Failures may be expected when the frontiers of knowledge are extended, as they were during the period of Gothic construction, or as they are today in the exploration of outer space. Unfortunately, many repetitive failures and accidents occur today not as the result of a lack of technical information, but as the result of procedural errors and information transfer deficiencies. These problems are discussed in Section 2.4.

Failures at the leading edge of technology have always been accompanied by periods of societal self-examination. Failures of symbolic works are especially vulnerable to questions that extend beyond the technical issues. The successive failures of the tower at the cathedral at Beauvais, France, created theological as well as technological questions (FitzSimons 1986). Similarly, the response to the Challenger space shuttle accident of 1986 involved questions and concerns far beyond the technical causes of the failure.

Henry Degenkolb has written that trial-and-error is the principal source of useful information for design when the input information is uncertain, as is the case for design of structures to resist earthquakes (Degenkolb 1980). Most of the Uniform Building Code provisions for seismic design are derived from observations of structural behavior during historic earthquakes. Degenkolb notes that the ultimate test of the adequacy of a mathematical approach to design is whether or not the theoretical solution agrees with the trial-and-error solution.

The design of mass-produced products often includes trial-and-

Figure 2.1 The Gothic cathedral builders exchanged failure and performance information during construction, while advancing the state of the art. (Courtesy T. Bartuska)

error as an integral part of the design process. Destructive or nondestructive testing of full-scale components or assemblies is routinely conducted in the design of aircraft, machine parts, and so forth. The luxury of destructive testing is usually not economically possible when designing a building or bridge. If the design is conventional, however, the designer does have the benefit of an inventory of past structures to consult for relevant performance information.

The design of some structures or products must be done without the benefit of trial-and-error experience. Some structures, because of their uniqueness, scale, or complexity, have no precedence. Examples include offshore platforms, large civil projects, or projects exposed to unusual environmental conditions. The design of such projects requires a conservative approach. All participants should be aware that frontiers are being extended. The situation calls for caution, such as that exercised by the Gothic cathedral builders.

James Amrhein, a designer of reinforced masonry structures, has defined "structural engineering":

> Structural Engineering is the art and science of molding Materials we do not fully understand; into Shapes we cannot precisely analyze; to resist Forces we cannot accurately predict; all in such a way that the society at large is given no reason to suspect the extent of our ignorance.

Perhaps this definition could be easily adapted to describe other fields of engineering design, given the uncertainties and professional judgments that must be made. Lessons learned from failures or accidents are extremely useful in clarifying some of the uncertainties, leading to improvements in the design process.

Communication among designers about lessons learned from failure has always been an important component in the advancement of the engineering professions. This process continues today. Lessons learned from experience are combined with mathematical theory to predict the behavior of engineered systems with a greater degree of confidence than ever before.

2.3 DEFINITION OF FAILURE

The discussion of failures, to be most productive, must include performance problems that are less than catastrophic. Many performance deficiencies are not life-threatening, but involve significant economic costs to society.

A comprehensive definition of "failure," as used by the Technical Council on Forensic Engineering of the American Society of Civil Engineers, is provided by Leonards (1982); "Failure is an unacceptable difference between expected and observed performance."

This definition is broad enough to include serviceability problems such as distress, excessive deformations, premature deterioration of materials, and inadequate environmental control systems. For example, in the case of buildings, the most costly recurring performance problems are those associated with building envelope performance. Leaking roofs and facades are not catastrophic, news-making events. But the forensic engineer spends a great deal of time investigating such problems. Dissemination of the results of these investigations can be of much economic benefit to the designers, owners, and operators of facilities.

Other industries have developed a vocabulary to encourage discussion of less spectacular performance problems. Dam designers and nuclear engineers prefer to use words such as *accident* and *incident* when discussing failures that are not catastrophic.

Performance of a structure or product must always be evaluated in reference to life expectancy and to the degree of maintenance provided (Figure 2.2). Gordon provides a context for evaluating performance of aging engineered facilities:

> The entire physical world is most properly regarded as a great energy system: an enormous marketplace in which one form of energy is forever being traded for another form according to set rules and values. That which is energetically advantageous is that which will sooner or later happen. In one sense a structure is a

Figure 2.2 Aging farmhouse structure near Lancaster, Pennsylvania. All structures are destroyed in the end.

Figure 2.3 Civic Center Coliseum failure, Hartford, Connecticut, 1978. Failure occurred 5 years after completion.

device which exists in order to delay some event which is energetically favored. it is energetically advantageous, for instance, for a weight to fall to the ground, for strain energy to be released, and so on. Sooner or later the weight *will* fall to the ground and the strain energy *will* be released; but it is the business of a structure to delay such events for a season, for a lifetime, or for thousands of years. All structures will be broken or destroyed in the end, just as all people will die in the end. It is the purpose of medicine and engineering to postpone these occurrences for a decent interval. [Gordon 1978]

Certainly, the forces of nature are sufficient to ensure that engineered facilities will not last forever. The work of the forensic engineer is to investigate those projects that do not provide the expected quality of performance for the expected period of time (Figure 2.3).

2.4   CAUSES OF FAILURE

Failures result from a variety of causes involving both technical/physical problems and human error/procedural factors (Khachaturian 1985). Often the root causes of a failure or accident are difficult to isolate and quantify. The causes may be a combination of interrelated deficiencies. This is one reason for the complexity and confusion that usually accompany failure litigation.

The underlying source of a failure is sometimes found in ignorance, incompetence, negligence and avarice, the "four horsemen of the engineering apocalypse" (FitzSimons 1986). Sometimes the project is exposed to environmental conditions or forces unpredicted by the designer. In the case of building or civil structures, the effects of natural hazards, such as fire, flood, extreme winds, or seismic events, may exceed reasonable predictions or accepted standards of practice. Even in these cases, however, carelessness in site selection, leading to unnecessary or reckless exposure to natural hazards, may be a factor.

When failures are discussed in professional journals, the typical article focuses on the technical/physical cause of the failure. There is a need for more discussion of procedural issues. There is always a technical/physical explanation for a failure, but the *reasons* failure occurs are often procedural. This is true even in the most advanced technologies, where formal quality control procedures are exercised. The Challenger space shuttle accident of January 28, 1986, was blamed partially on deficiencies in management of the quality control system (Figure 2.4). The Chernobyl, USSR, nuclear plant accident in April 1986 involved human error (Figure 2.5). The 1978 Willow Island, West Virginia, cooling tower collapse, which killed 51 construction workers, occurred as the result of procedural errors in construction sequencing (Lew 1980).

Dissemination of information about the procedural aspects of failures can help to affirm the role of the engineering professional and to improve standards of professional practice (Rubin and Banick 1987).

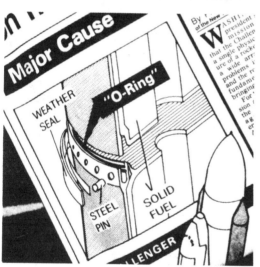

Figure 2.4 The Challenger space shuttle accident of January 1986 was blamed partially on mismanagement of quality control. There is always a technical cause associated with a failure, but the reasons are often procedural. (Courtesy *Lewiston Morning Tribune.*)

Procedural causes are usually interdisciplinary, involving communication deficiencies and unclear definition of responsibilities. Interdisciplinary information dissemination strategies are necessary to support the development of improved quality assurance/quality control programs.

For the purpose of discussion, the causes of structural failure in buildings and civil structures may be classified as follows (Carper 1987b).

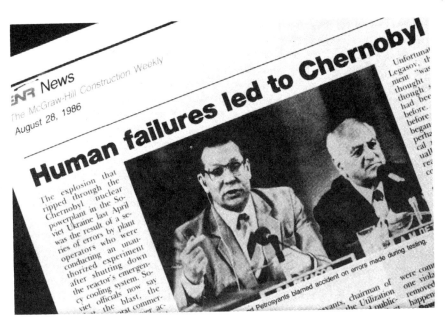

News
The McGraw-Hill Construction Weekly
August 28, 1986

Human failures led to Chernobyl

The explosion that ripped through the Chernobyl nuclear powerplant in the Soviet Ukraine last April was the result of a series of errors by plant operators who were conducting an unauthorized experiment after shutting down the reactor's emergency cooling system. Soviet officials now say the blast, the worst commer...

Figure 2.5 The nuclear accident at Chernobyl, USSR, in April 1986 involved human error. (Courtesy *Engineering News Record* and WideWorld Photo.)

A similar list could be developed for failures and accidents involving manufactured products.

1. *Site Selection and Site Development Errors.* Land-use planning errors, insufficient or nonexistent geotechnical studies, unnecessary exposure to natural hazards.
2. *Programming Deficiencies:* Unclear or conflicting client expectations, lack of clear definition of scope or intent of project.
3. *Design Errors.* Errors in concept, lack of redundancy, failure to consider a load or combination of loads, connection details, calculation errors, misuse of computer software, detailing problems including selection of incompatible materials or assemblies which are not constructable, failure to consider maintenance requirements and durability, inadequate or inconsistent specifications for materials or expected quality of work.
4. *Construction Errors.* Excavation and equipment accidents, improper sequencing, inadequate temporary support, excessive construction loads, premature removal of shoring and formwork, nonconformance to design intent.
5. *Material Deficiencies.* Material inconsistencies, premature deterioration, manufacturing or fabrication defects.
6. *Operational Errors.* Alterations to structure, change in use, inadequate maintenance.

A particular civil/structural forensic investigation may find a complex combination of several of the listed factors, all contributing to the failure. For example, failure of a hydroelectric project may involve mechanical and electrical problems, operational errors, design concept deficiencies, and geotechnical or structural inadequacies.

Failures resulting from corrosion in concrete parking structures have become commonplace in recent years (Robison 1986). Parking structures are among the most vulnerable constructed facilities. They are exposed to a wide range of fluctuating thermal conditions. Vehicles deposit corrosive deicing salts used on roadways onto the parking deck surfaces. These surfaces are not rinsed periodically by rain, as is the case for most highway structures. Parking facilities are utilitarian structures, low on the typical owner's maintenance priority list.

Investigation of a parking structure failure may uncover contributing factors from every category in the foregoing list. Design details may be inadequate, especially provision for drainage, detailing around openings, crack control, joint definition, and concrete cover. Specifications for sealant materials and application procedures may be insufficient. Geotechnical errors or foundation design problems may be present, leading to settlement cracking. Construction errors involving poorly executed connections or improper sealant application may have contributed to acceleration of corrosion. Sealant materials may be defective. The owner or operator of the facility may bear partial responsibility because of inadequate maintenance. Some owners have even used corrosive deicing salts in their own snow removal operations.

Sorting out an equitable distribution of responsibility for a costly failure in such a case is a formidable task. The use of multiple forensic experts, representing various parties to a dispute, can be helpful in ensuring that all potential contributing factors are considered.

2.5 DATA COLLECTION AND INFORMATION DISSEMINATION

The forensic engineer has an important role in assisting data collection and information dissemination efforts. The dissemination of accurate and complete information relative to forensic investigations can have an impact on improvements in engineering practice and products.

Many organizations are working to collect and disseminate failure and performance information (Carper 1987a). Most of these organizations have developed strategies to disseminate information among the practitioners in specialized fields. For example, the Association of Soil and Foundation Engineers (ASFE: The Association of Engineering Firms Practicing in the Geosciences) has published numerous case studies on geotechnical and foundation problems as part of the ASFE loss control program (Gnaedinger 1987).

The National Transportation Safety Board disseminates information instantly to policy-making bodies, including the Federal Aviation

Administration (FAA). Aircraft operating procedures are quickly revised as a result of these communications (Figure 2.6). Several of the contributors to this book have included references to dissemination activities within other specialized disciplines.

Recently, there have emerged some important data collection projects that have the goal of disseminating information on an interdisciplinary scale. In the previous section, it was noted that most of the procedural problems leading to failures and accidents are interdisciplinary in nature. Problems occur at the interface between specialists, where communications and responsibilities become confused. Efforts such as the Architecture and Engineering Performance Information Center (AEPIC) at the University of Maryland (Vannoy and Bell 1986; Loss 1987) and the Center for Excellence in Construction Safety at West Virginia University (Eck 1987) deserve the support of all engineering professionals. These projects seek ways to disseminate information to the most appropriate individuals and organizations, in the most appropriate form, so that lessons learned from accidents and failures can be most efficiently integrated into practice (Figure 2.7). Methods for introducing failure informa-

Figure 2.6 The results of an airline accident investigation by the National Transportation Safety Board are immediately disseminated to policy-making bodies. (Courtesy NTSB.)

Figure 2.7 Collapse of a roof structure during construction. (Courtesy Architecture and Engineering Performance Information Center.)

tion into the undergraduate engineering curriculum are under study. Forensic engineers have a responsibility to contribute to these projects.

The activities of the National Academy of Forensic Engineers also cross disciplinary lines (Section 1.4.1) Organizations such as NAFE present opportunities to address the interdisciplinary problems that have arisen as a result of specialization. The *Journal of the National Academy of Forensic Engineers* contains articles on a wide range of topics representing a variety of engineering disciplines.

In some fields, the collection and dissemination of failure and accident information can have an immediate and substantial impact. NTSB aviation accident information distribution policies have already been discussed. Operating procedures can be modified throughout the nation, and even throughout the world, as the result of NTSB recommendations. These changes in procedure can be implemented almost immediately after an accident (Chapter 7). The relationship between forensic investigation and public safety is direct.

In the case of defects in manufactured products, forensic investigations can also have a direct impact. Collection of failure and accident information can be used to support product recall efforts or to convince manufacturers to make necessary modifications to mass-produced products. Such modifications may affect a large number of similar products, having a significant immediate impact on public safety.

A building or civil structure is quite different. It is not a manufactured product. The design process involves a number of engineering disciplines and the result is a single, unique site-specific project. The rewards of performance data collection and information dissemination do not, on the surface, appear to be quite as direct as in the previous examples. Perhaps this is why data collection efforts for the construction industry have lagged behind those developed by other disciplines. There are, however, important opportunities within the construction industry for disseminating information about products, procedures, and assemblies. Many repetitive costly performance problems occur because of information transfer deficiencies. Information is needed to mitigate the problems associated with leaking roofs and facades in buildings. Problems with specific products and details occur over and over again, simply because informaion is not collected and disseminated to the appropriate decision makers.

For example, countless costly corrosion problems exist in buildings as the result of a particular mortar additive. Many users of the product were unaware of the potential problems even after litigation was in progress for some time. Many owners of facilities in which the product was used are still not aware of the deficient performance of this material.

Another example is the current volume of litigation associated with improperly designed and installed brick veneer/metal stud curtain wall panels. This assembly was heavily promoted and widely used, even as costly repairs were underway in facilities that first used the promoted system and deficient design guidelines. Even today, many designers are unaware of problems with this facade assembly. Clearly, the construction industry suffers from the lack of an established and coordinated information collection and dissemination program.

Several projects are underway within the construction industry to adress the need for interdisciplinary information dissemination. One of these is the *Journal of Performance of Constructed Facilities*, published by the American Society of Civil Engineers, Technical Council on Forensic Engineering. The quarterly journal, established in 1987, is also sponsored by the National Society of Professional Engineers and the Architecture and Engineering Performance Information Center. An interdisciplinary editorial board reviews papers for publication on the causes and costs of failures and other performance problems in constructed facilities. The journal is open to discussion by all professionals involved in any aspect of the construction industry.

2.6   FAILURE TRENDS AND PROFESSIONAL RESPONSE

2.6.1   Trends Leading to an Increase in the Frequency and Severity of Failures

Today's technological society is increasingly vulnerable to the effects of failures and accidents. Dense urban population centers have aggravated the potential effects of natural and human-caused disasters.

Centralized utilities and industrial facilities are extremely vulnerable to the effects of human error or equipment failure. An error made by a single individual can have catastrophic effects. For example, a construction worker recently dropped a socket wrench while working on a power plant project in Florida. The dropped wrench started a chain reaction of events that culminated in $5 million worth of damages to the facility and put the entire plant out of operation for 4 months.

Human error or equipment failure in the operation of a large aircraft or ocean vessel can result in the loss of hundreds of lives in a single accident. A single connection failure in a long-span structure can threaten the lives of thousands of occupants. A significant industrial accident, such as the methyl isocyanate gas leak experienced in Bhopal, India, in December 1984, can cause tens of thousands of casualties. The concentration of hazardous wastes near population centers is of great concern, as is the transportation of hazardous materials through populated areas.

Along with the increasing vulnerability to severe effects of accidents, some alarming trends have appeared in the attitudes and expectations of society with respect to engineered products and facilities. These trends could lead to an increase in the frequency of failures unless the engineering professions develop appropriate mitigating strategies.

The American Society of Civil Engineers, Technical Council on Forensic Engineering organized a workshop on "Reducing Failures in Engineered Facilities" in Clearwater, Florida, in January 1985 (Khachaturian 1985). This workshop was attended by professionals representing all interests in the construction industry. Wide-ranging discussions were held regarding the causes and costs of failures and accidents in the construction industry. The general consensus that emerged from this conference was that both the frequency and severity of failures were on the increase. There was not, however, general agreement as to the reasons for this increase. Several contributing trends were noted during the discussions and are included here. Some of these issues are unique to the construction industry, but others are present, to varying degrees, in all fields of engineering practice.

One problem noted by many practicing professionals is the loss of control by the designer over execution of the design. This loss of control is the result of an undue emphasis on reduction of construction time and construction cost. Nontraditional "fast-track" approaches have been introduced, with new opportunities for misunderstood communications and unclear lines of responsibility. Designers have always had to compete against the costs of time, but the unprecedented era of double-digit inflation in the 1970s placed new pressures on design and construction schedules. Society began to define the "best" project as the one that could be completed in the shortest time, for the least initial cost. The "psychology of the low bid" became fully entrenched in the construction industry. Competitive bidding for design services and the ethical conflicts of design–build package services have become potential contributors to escalating failure rates.

Along with this emphasis on reducing costs and construction time has come the demand from society for projects of ever-increasing scale and complexity. The process of designing and constructing complex modern buildings has necessitated the development of a large number of engineering specialties. Coordinating the work of these specialists has become an increasingly difficult task, and communication problems have escalated with the number of participants. Specialization requires a greater emphasis on coordination. Problems are developing at the interface between the work of the various specialists.

New materials and methods are appearing at an unprecedented rate. These materials and procedures are promoted and applied simultaneously in a variety of environmental exposures without the benefit of long-term trial-and-error experience.

The computer has been introduced into the construction industry only recently. The potential exists for the computer to contribute toward resolving some of the current problems within the industry; however, problems are also occurring as the result of misapplication of this new tool. The computer can process information very rapidly, but the time for reflection by the designer may be lost. The computer can be used to explore a wide range of alternatives. It can be used to relieve the designer of repetitive tasks, enabling the designer to devote more time to quality control, including increased participation in the construction process. Unfortunately, many have seen the computer as a tool that can merely speed up the design process.

The availability and proliferation of specialized software are also of concern. The probability exists that software will be misused by unqualified designers, as an alternative to consulting specialized professionals who are qualified to make the necessary experience-based engineering judgments.

Expanding litigation has ironically contributed to the increasing frequency of failures. One would expect that the threat of litigation would cause all participants to exercise a greater level of care in their duties. The *uncertainty* of the outcome of litigation, however, has caused many designers to see litigation as simply a "cost of doing business," an inevitable aspect of engineering practice. As a result, more time is spent on writing contracts to limit liability, leaving less time to devote to engineering, inspection, and quality control. The resulting "tighter" contracts also unfortunately may leave certain responsibilities unassigned.

In response to expanding litigation and mistrust of the outcome of litigation, many designers have developed a reliance on insurance. It has been suggested that *insurance* has been substituted for *quality* in the construction industry. Such is not in the interest of public safety.

Forensic engineers can make a substantial contribution toward counteracting or reversing some of these trends. In particular, the presentation of sound engineering arguments to effect rational legal decisions can help to restore faith in the legal system. Other strategies are also emerging. Many forensic engineers are actively working within professional societies to respond to the listed concerns.

2.6.2   Failure-Reduction Strategies

Whenever conditions have arisen that impact public safety, the community of professional engineers has been among the first to show concern. At the present time, many projects are underway to respond to the negative effects of the foregoing trends. These failure-reduction strategies are supported by a broad base of individuals and professional societies.

Interdisciplinary conferences have been created such as the workshop on *Reducing Failures of Engineered Facilities* (Khachaturian 1985) and the Engineering Foundation conferences in Santa Barbara, California, and Palm Coast, Florida, on "Structural Failures." These events, attended by professionals representing a wide range of disciplines and interests, present unique opportunities for exchange of information. Many of the critical procedural deficiencies resulting from specialization can be resolved through open communication between disciplines.

Projects are underway that seek to more clearly define the roles and responsibilities of the various parties in the construction industry. Unclear lines of responsibility have in the past contributed to many failures and accidents.

New methods of dispute settlement are proving to be more successful than traditional litigation, in certain cases involving engineering failures (see Chapter 13). Innovative approaches to insurance, which encourage cooperation rather than confrontation, are emerging. Many of these insurance programs are coupled with loss-prevention incentives.

The computer is being used successfully to assist in the management of procedures, and to cross-check the coordination of design and production. Also, computer methods are now used to better coordinate the collection, collation, and dissemination of performance information, through such projects as the Architecture and Engineering Performance Information Center.

Figure 2.8 Houston, Texas. Dissemination of information regarding performance of curtain wall systems during Hurricane Alicia (1983) will contribute to improved design and construction practices.

Quality assurance/quality control programs, including peer review, enhanced communication techniques, and greater participation by the designer in the construction and manufacturing processes are being tested and refined. Peer review programs have been endorsed by many professional societies including ACEC, ASCE, and ASFE.

There is more discussion than ever before in the current engineering literature about failures, accidents, and performance information. This exchange of information has brought about improvements in performance (Figure 2.8).

Forensic engineers have been very much involved in suggesting and implementing all of the foregoing failure-reduction strategies. Opportunities exist for an expansion of these activities in the interest of mitigating the frequency and severity of failures.

2.6.3 Summary

As long as structures and products are designed, built, and manufactured by humans using imperfect materials and procedures, failures will be experienced along with successes. The tools and theories of modern technology are available to enable the designer to do a better job than ever before in ensuring quality in the constructed project or manufactured product.

Unfortunately, there are trends in modern society that may lead to an increase in both the frequency and the severity of failures. The engineering professions have recognized these trends and are developing strategies to mitigate their effects.

It is expected that the role of the forensic engineer in contributing to improved practices will become more visible in the future. Data collection and information dissemination activities will expand to support the development of effective quality assurance/quality control procedures. The engineering community will continue to learn from experience, following the time-honored trial-and-error tradition.

References

Carper, K. L. ed. 1986. *Forensic Engineering: Learning from Failures*, New York: ASCE.

———— 1987a. Failure information: Dissemination strategies, *Journal of Performance of Constructed Facilities* (ASCE) 1, No. 1 (February).

———— 1987b. Structural failures during construction, *Journal of Performance of Constructed Facilities* (ASCE) 1, No. 3 (August).

Degenkolb, H. J. 1980. Failures: How engineers learn—Or do they? ASCE Convention Paper, Portland, OR, April 14–18, Paper Preprint No. 80-017, ASCE, New York.

Eck, R. 1987. Center for Excellence in Construction Safety, *Journal of Performance of Constructed Facilities* (ASCE) 1, No. 3 (August).

Feld, J. 1964. *Lessons from Failures of Concrete Structures*. Detroit, MI: American Concrete Institute.

———— 1968. *Construction Failure*. New York: John Wiley and Sons.

FitzSimons, N. 1986. An historical perspective of failures of civil engineering works. In *Forensic Engineering: Learning from Failures*. New York: ASCE.

Gnaedinger, J. P. 1987. Case histories—Learning from our mistakes. *Journal of Performance of Constructed Facilities* (ASCE) 1, No. 1 (February).

Godfrey, E. 1924. *Engineering Failures and Their Lessons.* Akron, OH: Superior Printing Co.

Gordon, J. 1978. *Structures: Or Why Things Don't Fall Down.* New York: Penguin Books.

Khachaturian, N. ed. 1985. *Reducing Failures of Engineered Facilities.* New York: ASCE.

Leonards, G. A. 1982. Investigation of structural failures. *Journal of the Geotechnical Engineering Division* (ASCE) 108, No. GT2 (February).

Lew, H.S., 1980. West Virginia cooling tower collapse caused by inadequate concrete strength. *Civil Engineering* (ASCE) 50, No. 2 (February).

Loss, J. 1987. AEPIC Project: Update. *Journal of Performance of Constructed Facilities* (ASCE) 1, No. 1 (February).

Mark, R. 1982. *Experiments in Gothic Structure.* Cambridge, MA: MIT Press.

Petroski, H. 1985. *To Engineer Is Human.* New York: St. Martin's Press.

Robison, R. 1986. The parking problem. *Civil Engineering* (ASCE) 56, No. 3 (March).

Rubin, R. A., and L. A. Banick. 1987. The Hyatt decision: One view. *Journal of Performance of Constructed Facilities* (ASCE) 1, No. 3 (August).

Vannoy, D. W., and G. R. Bell. 1986. Data collection and information dissemination. In *Forensic Engineering: Learning from Failures.* New York: ASCE.

3. Fire Investigation

PAUL E. PRITZKER, P.E.

3.1 INTRODUCTION

This chapter is an introduction to the technique and training required by the engineering team to obtain fire investigative results of merit. Typical assignments are presented in a brief format illustrative of the range of cases and of the focus required for those who are professionals in this field.

This chapter does not dwell on the comprehensive body of knowledge that is prerequisite to becoming a qualified field fire investigator. Fire investigation requires knowledge of fire spread and behavior of fire, as well as an understanding of ignition sources and causation of fires and explosions. One should have a working knowledge of fire codes and standards and an affinity for a lifelong commitment to learning, as an adjunct to prior knowledge.

3.2 BACKGROUND

Whoever coined the term *working fire* was a master of understatement. Fire fighting ranks high on virtually every list of hazardous occupations.

The effort to determine the probable cause of a fire is to some degree an art and, to an even greater degree, a science. Notwithstanding that fires frequently destroy significant quantities of physical evidence, fire investigation can be productive when forensic engineering techniques are utilized. One has to exercise care in investigating a fire. Loose debris, open holes, and weak structural components can place the fire scene investigator at risk. But the investigator must deal with these risks, as he or she "digs" for the evidence to support analysis and the formation of a conclusion that can be defended in an adversarial procedure.

Fire investigators do not all focus on the same issues. A fire investigator who is a fire department employee, working alone or as part of an "arson squad," has a primary focus on establishing the probable cause of the fire. In most situations an effort is made to determine if there was arson for profit. In areas having a high incidence of incendiary fires, the arson squad will try to determine clues that might connect this fire with other fires currently under investigation. "Fire investigation" is more properly referred to as "fire and explosion investigation." It is critical to determine the cause and origin of a fire or explosion, whether the incident was accidental, incendiary, or of natural origin.

Not infrequently, fire investigation assignments are made after the cause and origin of the fire have already been determined. A fire in a new hotel under construction in New York led to an analysis of building codes and standards to ascertain whether current code-mandated procedures provided "adequate" life safety protection during construction. A fire in a large hotel in Nevada led to an analysis of smoke transport to the upper floors.

Many of these assignments use the services of multiple practitioners. These may include National Fire Protection Association (NFPA) teams and National Bureau of Standards (NBS) experts, who publish their reports for use as technical resources. Private forensic engineers report in confidence to their own clients, most of the time.

3.3 INVESTIGATIVE TECHNIQUES, PROCEDURES AND TOOLS

Each fire investigation is unique, but a graphic description of the typical sequence of an investigation is given in Figure 3.1. Following are some observations regarding techniques, procedures, and tools used in typical fire investigations.

One would assign a police officer to traffic duty before sending him or her on a homicide investigation. One would expect to serve as a junior member of a homicide investigation team for an extended period of time before developing the expertise to assume leadership responsibility for an assignment of significant magnitude. This caveat is no less true for the field of fire investigation.

Years ago, when I was a junior member of a fire analysis investigative team, I was taught an adage: "Judgment comes from experience; experience comes from bad judgment." One of the most difficult tasks is to come to the site of an investigation without preconceived ideas as to the cause, origin, or responsibility for the fire that is under investiga-

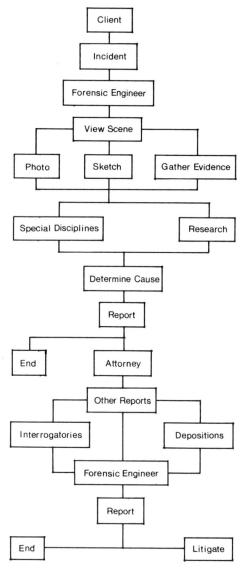

Figure 3.1 Path of a typical assignment.

tion. A forensic investigator must always keep an open mind and must avoid the tendency to look for facts to fit a preconceived notion.

Clients may contribute to this problem if they try to direct the focus of an investigation. I recall an incident in which the investigator was asked to pay special attention to a copy machine in an investigation involving an office occupied by a member of Congress. The machine was suspected to have overheated and ignited a nearby combustible source. After a prodigious amount of time was spent inspecting the copy machine, it was concluded from the physical evidence that the locus of the fire was remote from the copy machine. The cause of the fire was eventually attributed to failure of a secretary to turn off a hot plate. Had the fire started as the result of a defective copy machine, the insurer would have had an opportunity to recover via subrogation against the copy machine manufacturer. Because the cause was determined to be employee human error, no such recourse was available.

Wellington once said that the Battle of Waterloo was won on the playing fields of Eaton. It is important to prepare before arrival at the fire scene. When possible, photographs of the facility and drawings and/or specifications prepared by design professionals should be reviewed. Frequently, with the assistance of an owner or other informed person, accurate diagrams can be constructed when formal drawings are unavailable. Prior to the site visit, official fire reports prepared by local fire department investigators and by the state fire marshal should be obtained and reviewed. This background information allows the investigator to be more efficient in on-site field inspection.

When the fire investigation includes an analysis of the contents of the building, an "official list" of equipment alleged to have been on the property at the time of the fire should be obtained prior to the site visit. The list may be obtained from the client. Such a list is required as part of the proof of loss for processing claims. During the site visit, location of the equipment should be verified. It is good practice to record the model and serial numbers of all equipment as a standard procedure for inventory documentation.

One cannot take too many notes, either written or on a small portable tape cassette recorder. If a recorder is used, spare tapes and batteries should always be available in the field. As a standard procedure, I always note all items in each room and play back the recording to make certain the notations were recorded, before moving on to the next room or area.

Photographic documentation is critical in professional fire investigations. Chapter 10 contains much useful information regarding the use of photography in forensic investigations. The following are some comments relative to the specific use of photography in documenting a fire scene.

It has been said that a good photograph is "tantamount to stopping the clock." Because trials may occur years after an event, a good photograph is essential when job-site memory fades. During the various phases of digging for evidence, photographs that record the various intervals enhance credibility. There are many excellent books that address in great detail the topics of shutter speed, closeup (macro) lenses, wide-angle lenses, and telephoto and zoom lenses. These references should be in every serious professional fire investigator's library. In general, the latest generation state-of-the-art 35-mm automatic flash cameras are compact and adequate for the task of obtaining high-quality photographic documentation.

The conditions observed and preserved on color photographic film are vital to future analysis. The use of color film cannot be overemphasized in the quest to compare burn patterns and debris components.

One can attempt to determine the substance of burning material from a study of the color of the smoke and flame. Burning wood produces gray to brown smoke with a flame color of yellow to red. Gasoline produces black smoke with a yellow flame color. Such distinctions are not possible to record with black-and-white film.

Videotapes are effective, efficient, and portable. Use of videotaping as an adjunct source of documentation helps to obtain an overview.

For precise photographs to use in a report, individual 35-mm color photographs provide an invaluable record. The use of instant photography is also popular and effective. Instant photographs are best used for instant documentation to verify that the image of the evidence is captured. The new accelerated one-hour photo processing has, to some extent, diminished the special advantage of instant photography. As a standard procedure on out-of-town assignments, I search for a local one-hour photo laboratory before leaving town to make sure I have captured the significant images on film.

Film is inexpensive. The investigator should not be frugal in the use of film. It is no sin to take too many photographs. There is a potential problem in taking too few.

A system of organization is important in photographing a fire scene. First, the investigator should record an exterior view of the neighborhood. All sides of the property should be documented on film. This thorough procedure should also be followed for interior photographs. One cannot be positive of the probable cause of a fire unless all items are documented, including those that did not contribute to the cause of the fire. Every power apparatus and heat-producing item should always be photographed. These include gas, oil, and electric systems. These photographs provide evidence that these items were examined by the investigator, and for reasons noted, they were eliminated from the list of probable sources of attribution for the fire cause.

Before the investigator goes to the fire scene, arrangements must be made to ensure a safe environment for the investigator. Someone in responsible charge should be contacted to arrange for safe access. Structural shoring should be provided before entering a fire scene. Adequate lighting must be provided for safety and efficiency. Proper clothing and protective eyewear and footwear are essential. A fire investigator is not performing an Olympic event. One should never compromise personal safety.

Fire investigation as a team is usually more effective and less hazardous. There is safety in numbers. Among the hazards commonly encountered by the fire investigator are water in basements and nails protruding from boards. In entering a fire scene, *CAUTION* is the key word.

Time is an enemy to the fire investigator. When one is called late to investigate a fire, physical evidence may have been altered or may have deteriorated. This may preclude an analysis of merit. In the course of fighting a working fire, damage frequently obliterates clues that might have been beneficial to the task of determining the locus and source of ignition.

When a request is received to investigate a fire, it is critical to confer with the local fire department officials and to interview the chief or the chief's deputy, if possible, to ascertain who was on the scene. Copies of all fire reports should be obtained. A fire investigator is most effective when a unique symbiotic relationship is established with the members of the fire service.

3.4 TRAINING

Fire investigation is a comprehensive task that requires the blending of a variety of technical skills. Various methods have been utilized for training fire and explosion investigators. These include formal university education programs, correspondence courses, seminars, short courses, and "on-the-job training."

Fire investigation based only on college-level classroom training is often taught from textbooks by faculty who have not had practical experience. Academic qualification devoid of practical experience does not provide the depth of expertise necessary to educate one who must search debris for technical clues. A pristine ivory tower is a quantum distance from the rubble left in the aftermath of a major conflagration. There are, of course, situations that require skills developed in a college education. When one is investigating a fire of suspected electrical origin, an electrical engineering background is critical to the analysis. Computer modeling techniques, knowledge of chemical properties, affinity with the concept of flash points, and other similar relationships can be learned in the classroom. These concepts are all used by forensic engineers in fire investigations.

Traditional educators have often denigrated correspondence course education to the hinterlands. I submit, however, in training for fire and explosion investigation, that an astute student who completes a recognized correspondence course will obtain a substantial depth of knowledge. This method has the advantage of learning at one's own pace. Another feature of correspondence education is that it allows the fire investigator to continue to work as a junior member of a team while training is taking place. There is at least one outstanding course that consists of twelve complete and very detailed lesson plans, culminating in a final written examination.

I am a member of a National Board of Certification for Fire and Explosion Investigators. One technique I have advocated is a telephone interview with the student, subsequent to grading the final examination. Other board members have asked the local fire chief to interview the student prior to granting the final grade for the course and certification. In one group, nearly 70 percent of the certifying students had some college training, at least to the level of an associate degree.

Short courses and seminars are useful for continuing professional development. Professional education must be a lifetime commitment if one is to maintain technical competence.

Without a sound background in the art and science of fire and explosion investigation, a report on the cause and origin of a fire may not be worth the paper on which it is written. An effective investigator must have an understanding of the interplay of chemistry, physics, fire protection methodology, and forensic engineering.

Arson is responsible for less than 15 percent of residential fires, less than 30 percent of commercial fires, and less than 25 percent of industrial fires. These statistics show that persons who have training limited to arson investigation are not sufficiently qualified to conduct successful investigations on the great majority of fires that occur in this country daily. There are many important topics that are not always included in arson seminars, for example, spread of fire; explosions, including liquid petroleum and natural gas, dust, and boiling liquid/ expanding vapor; electrical causes of fire; static electricity; lightning fires; and careless disposal of combustible material (i.e., cigarette smoking in bed). Arson investigation training is important, but one should be aware that fire investigation requires a much more comprehensive degree of training and expertise. Arson investigation training has merit, but it must be augmented by further education.

3.5 THE ENGINEERING TEAM

In the delivery of professional services, seeking a second opinion is considered prudent. Many of us seek a consultation with another surgeon when our physician has advised surgery. Attorneys and certified public accountants often consult colleagues to assure they are providing optimal advice. Fire investigators often do likewise.

As previously discussed, fire investigation is frequently a multi-disciplinary activity. Most fire investigators are not electrical engineers. In many cases, that special expertise is not required. Electrical engineers are used as consultants to determine whether a fire was of electrical origin. They have the special expertise to address the concept of the electrical circuit and the conditions necessary to cause a fire.

There is a corollary requirement for utilizing the services of a consulting electrical engineer. Sometimes, members of the fire service are reluctant to admit that they do not know the cause of a fire. Instead of listing the cause as "undetermined," they may attribute it to "electrical causes." Of course, the term *Electrical Fire* is a misnomer. Electricity does not burn. Electricity is a form of energy that can be converted into heat.

To cause a fire of electrical origin, electricity must have a path to the place where it is converted into heat. The heat must rise to such a level that the kindling temperatures of adjacent flammable materials are reached, causing ignition. The analysis of conductors, insulators, switches, devices, fuses, circuit breakers, resistance, and arcing is the province of the electrical forensic engineering fire investigator.

An electrical engineer understands that electricity can be changed to heat and light energy when it passes through an ionized gas. Arcs produce extreme temperatures that are localized, but these temperatures can spread if overcurrent devices do not function properly. A temperature rise in excess of 5000°F can melt aluminum, which fuses at 1200°F, or copper conductors, which fuse at 1980°F, or steel enclosures, which fuse at 2200°F.

The presence of "beaded wires" indicates that arcing has taken place. Beaded wires are produced when the melted ends of conductors resolidify, forming a small spherical globule of metal called a "bead." Beading, by itself, indicates only that arcing has occurred. It does not prove that arcing was the cause of the fire.

To be an effective contributor to the investigative team, the electrical engineer must have a working knowledge of the *National Electrical Code* (NFPA-70) published by the National Fire Protection Association, Quincy, Massachusetts.

Another important member of the engineering team is the chemical engineer. Chemical fires include fires involving chemicals in the ignition, cause, or spread of the fire. Chemicals include any substance or combination of substances not normally expected to be in the environment under analysis.

Spontaneous combustion can result from the interaction of various chemical combinations. A chemical may serve either as a fuel or as an accelerant. Release of oxygen increases the fuel content, resulting in a more intense fire. This leads to higher temperatures and acclerates the fire spread.

Frequently, a metallurgist is utilized as a member of the engineering team. When aluminum conductors appear to have overheated as a result of improper connection techniques, when "beading" is observed, or when components have failed, the technical interface with a metallurgist can be essential.

Mechanical engineers play an important role as members of the fire investigation engineering team. They may investigate machine parts, such as bearings or other sources of friction. On boilers and HVAC systems, they focus on technical nuances unique to their area of expertise.

Special discipline expertise is required when cutting and welding fires are investigated. Vehicle, aircraft, and boating fires require investigators who are familiar with all of the systems endemic to these entities. Dust explosions, which are extremely lethal, require special hazard expertise. A specialist may be needed to verify whether or not a natural gas or liquid petroleum line or fitting leak caused a fire.

One advantage of the team approach is the collaboration it provides to back up courtroom testimony. Not all investigations lead to a trial, of course, but sometimes a trial takes place long after an investigation. The presence of a redundant site investigator can add insurance, in case the principal investigator is no longer available.

It is not always possible to bring in a complete team on the first day of an investigation. After the initial site inspection, one or more follow-up visits can be productive. The team members must be able to work well together. They must enjoy field work. They must have expertise and patience.

3.6   FIRE INVESTIGATIVE RESULTS

The initial focus of a fire investigation is to determine, with reasonable forensic engineering certainty, the probable cause of the fire. In point of fact, there are other long-range benefits that can result from that effort.

Changes are made in product design when manufacturers learn of the consequences of use of their product in the marketplace. Changes have resulted, for example, in the treatment of children's sleepwear. Changes in the control design of process equipment have followed from recommendations contained in fire investigative reports.

Many of the significant changes in the model building codes have resulted from the tragic loss of lives in fires and the subsequent investigations. Among the more influential fires were the 1942 Coconut Grove Fire in Boston, Massachusetts, and the 1980 MGM Grand Hotel Fire in Las Vegas, Nevada.

Automobile batteries can explode under foreseeable conditions. Several design changes and warning modifications have resulted from field investigations and forensic engineering reports involving these incidents.

Because the large majority of fires are accidental, those who report on the causes of fires and explosions alert the nation to danger and serve as a catalyst for change.

3.7   FIRE INVESTIGATIONS: CASE STUDIES

Several case studies, and the lessons learned, are included here to illustrate the wide range of activities involved in forensic fire investigation. These case studies are from the writer's own files. Readers who desire more detail regarding any of these cases are encouraged to contact the writer. The case study should be identified by the indicated file number.

3.7.1 High-Rise Apartment Incinerator Door (No. 21160)

A tenant in a high-rise apartment building opened the incinerator door in the hallway adjacent to her apartment. She alleged that she was burned when she attempted to insert trash into the chute. There were no witnesses.

A site visit included an inspection of the incinerator located in the basement of the building. The fire box was 8 feet wide with a slotted grate for ash to drop into a reservoir. An 18-inch hinged door allowed the building superintendent access to the burning chamber. A blower and flue gate operated on a 20-minute timed cycle for a sequence period of 2 minutes. Adjacent to the fire box a scrubber met Environmental Protection Agency (EPA) and state and local clean air codes. The system was fueled by natural gas. Timers sequenced three sustained burning periods.

Each of the six upper floors contained a hinged counterbalanced door that deposited trash via gravity into the incinerator. Above each door, prominently displayed, was a sign with graphic illustrations informing tenants of items that should not be thrown down the chute.

The probable cause of the injury was the explosion of an aerosal can, causing a flashback to the second floor.

3.7.2 Industrial Equipment Fire (No. 20881)

An employee was feeding aluminum/teflon-coated pieces of stock into the input end of an automatic sanding belt conveyor. A fellow employee was removing the product from the discharge end of the conveyor. A fire ensued. The plant was engulfed with dense smoke, making it difficult for the employees to find their way out of the factory.

The sander was connected to a portable dust collector located in a 55-gallon drum. The sander contained safety devices to eliminate contact with the metal frame by the belt, in the event the belt became untracked. This was intended to prevent sparks.

A service technician had noted a short time prior to the fire that the machine was in poor condition and had not been serviced for 3 years. The owner of the industrial facility did not authorize the recommended repairs. The three-phase electric motor was connected in reverse rotation. A dozen unsafe practices were documented. Violations of NFPA-63 (Dust Explosions), NFPA-65 (Processing/Finishing of Aluminum), and NFPA-70 (Electrical Code: Hazardous Locations) were detected. Our investigation concluded that the plant management was cavalier in violation of NFPA codes to the degree they were playing "Russian roulette." The cause of the fire and explosion was inept management.

3.7.3 Condominium Natural Gas Explosion (No. 20292)

It is difficult to understand why someone who does not have time to do something right the first time can find the time to make it right the second time.

A young man purchased a condominium and asked the utility company to turn on the gas. He left his keys with the building superintendent. When the gas utility serviceman arrived at the property, he turned on the gas meter. He lit the gas furnace and the water heater next to the furnace.

After igniting the basement gas appliances, the serviceman and the superintendent approached the top of the basement stairs on their way to the kitchen, where they smelled gas. The gas utility serviceman turned the stove valve off, checked the wall thermostat, and left the premises. In a short time an explosion and fire ensued.

An inspection of the facility was conducted. Surprise! The gas line serving the new stove was never connected. It was a direct open gas line that had no plug on the end. The person who removed the old stove was negligent in failing to cap a live gas line. The utility gas serviceman was also culpable because he turned the gas meter on and ignited the two gas appliances in the basement before he went upstairs. He did not notice the open gas line to the stove. He did the second time.

3.7.4 Cooking Fire: Improper Stove Part (No. 20383)

The young married couple rented an apartment in a suburban village. The young groom enjoyed cooking. One Sunday morning, he decided to surprise his bride with his culinary expertise: French fries. In a short period of time, she got the surprise of her life.

He placed a shallow pan filled with cooking oil on an electric burner. As he began to cook the French fries, the burner popped out of place, causing the hot oil to spill onto the burner. This resulted in a very hot flash fire. The man was cooking without a shirt, and apparently some of the fire burned his chest immediately upon acceleration.

He attempted to take the burning pan of oil out of the kitchen to protect his wife who was standing next to him. As a result of this attempt, more hot oil spilled on his chest and legs, causing severe burns to both. The carpet also ignited.

No one should ever stand over a hot stove without adequate clothing. Notwithstanding this imprudent attire, it was determined that the burner had a history of "popping out." Prior to the young couple's lease, the original stove pan had been replaced with a "Brand X-universal" stove pan. The bottom was square instead of round and it did not fit properly. The improper stove part was a major cause of this tragic fire.

3.7.5   Malfunctioning Television Set (No. 19465)

Some cases are so tragic that a bifurcated analysis is desirable. A small bedroom (100 square feet) was occupied by a young lad who suffered extremely severe third-degree burns. He was engulfed in a conflagration as he slept in his bed.

Before focusing on the probable cause of the fire, a documentation of all factors that could possibly have contributed to the fire was undertaken. A scale drawing of the building was prepared. All of the electrical outlets, branch circuits, and overcurrent devices were located, and it was determined that none of these played a part in causing the fire.

The washer, dryer, cooking equipment, heating equipment, and telephone systems were examined and found not to have contributed to the fire. There was conclusive evidence that the young lad was not playing with matches. He was not smoking.

When all of these items were eliminated by this "diagnosis of exclusion," there were strong clues that pointed to the television set located in the boy's bedroom. Two consultants were retained to buttress the team. In a laboratory, a comprehensive autopsy of the television set was documented in a methodical process (Figures 3.2(a) and (b)). A fire expert from out of town made an independent analysis of the investigation. The television set malfunctioned. A young man's life was destroyed.

3.7.6   Electrical Accident: Governmental Building (No. 20173)

State employees were given an unscheduled holiday when a short circuit on the State House switchboard caused an explosion, prodigious smoke, and a small fire.

The local electric utility company apparently was concerned that a proposed fire pump scheduled to maintain pressure on the stand pipes might have been connected to the line side of the West Wing 4000-amp 277/480-volt three-phase four-wire wye line ahead of the company's meter. Beyond periodic testing, the pump was not scheduled to stay on line.

Without prior notice, the utility company's employees entered the power room and began work on the live switchgear to relocate the potential connection. In the process a tool grounded to the bus. A flash and a loud bang sent the crew scurrying for safety. The lineman thought his clothes were on fire. He felt his eyes burning.

When the investigation was conducted, contact burns on a screwdriver were detected (Figure 3.3). The Electrical Code (234-94 Exception # 4) allows the supply to fire pump equipment to be connected on the supply side of the overcurrent device. The work on the switchboard could have been scheduled for times when the facility was shut down. The rush to save a few future pennies caused an immediate "megabuck" loss.

a

Figure 3.2 (a) Technical expert inspecting the television chassis. (b) Technical expert inspecting the rear section of the television.

b

Figure 3.3 Section of switchboard bus grounded by contact with tool.

3.7.7 Vacation Home: Inadequate Egress (No. 20255)

An owner of a distributing company allowed family, employees and special customers to vacation in his large mountain retreat. The wood frame 48 × 48-foot house was located on a remote site in a mountainous area. The second-floor bedrooms, one in each quadrant, overlooked an atrium that contained a fireplace. (see Figures 3.4(a)–(c).)

The guests went to bed quite late (early morning). A smoke alarm activated. The heat was too intense and the smoke too dense to allow for safe egress. The single wooden stairway was consumed in flame. One male guest broke a window. He and his wife were injured from the fall to the ground. Another male guest left his wife, went to his children's bedroom, and threw the small children outside into the arms of his brother-in-law. His wife and his parents died in the fire. Their charred bodies were found in the rubble after the fire burned the home to the ground.

A comprehensive analysis ruled out several potential causes of the fire including arson, smoking, the coffee pot, the fireplace, and defects in any utility systems. The precise cause of the fire may still be subject to debate. The loss of life, however, was caused by foreseeable factors indicated in the figures. The inadequate egress from the bedrooms, the removal of the balconies from the original design, the narrow glass

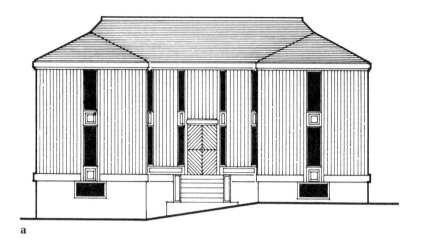

a

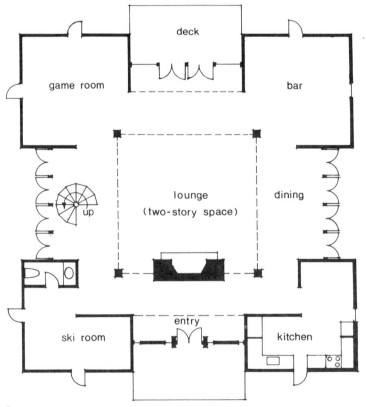

b

windows, and the single narrow winding stairway exacerbated the potential for loss of life.

An owner who allows infrequent visitors to use a property for a vacation should be considered similar to an innkeeper. Both have a duty to provide adequate life safety features. Remote cabins can have rope ladders, balconies, fire extinguishers, and doors, rather than narrow windows, for the purpose of rapid egress.

Fire investigations are beneficial when they can determine the probable cause of the fire. They are also beneficial when they articulate life

Figure 3.4 (a) Elevation of vacation home. The narrow window openings contributed to the difficulty of egress. (b) Main-floor plan. Note the single narrow winding wooden stairway to the second-floor sleeping area. (c) Second-floor plan. The original design provided for a swinging door to a balcony for egress from each of the four bedrooms. The balconies were eliminated in the final construction.

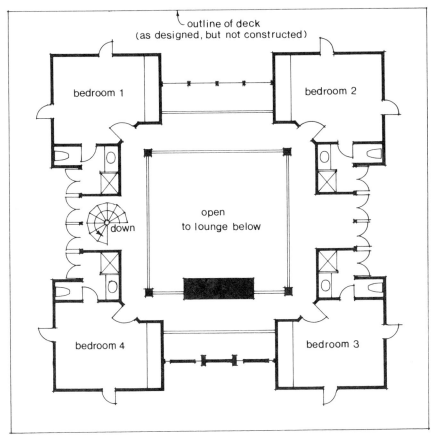

c

a

b

safety deficiencies that must be addressed so that future tragedies can be prevented. To paraphrase George Santayana: Those who fail to heed the lessons of history are condemned to repeat them. A professional forensic fire investigator can communicate findings in such a way that needless tragedies are not repeated.

3.7.8 Boating Accident (No. 20963)

It was an early spring. The doctor and his wife put their boat in the water for the start of a fresh season at the oceanside marina. They went to the fuel dock, filled the gas tank, and started the engine, without taking adequate time to ventilate the bilge. An explosion burned the boat to the water line. (See Figures 3.5(a)–(c).)

The captain and his mate escaped with minor singes. To them, however, the term *pleasure boating* became a misnomer.

c

Figure 3.5 (a) View of port side of boat, showing minimal damage. (b) Starboard side, showing massive damage. (c) Another view, starboard side of the boat.

3.7.9 Arson (No. 20738)

During an inspection after a residential fire, it was observed that the springs of a sofa were flattened, and that the wall was discolored by burning above the wainscot (Figure 3.6). The flat spring is a consequence of high temperature that has removed the tempering from the spring. This condition is frequently indicative of a fire caused by smoldering of a cigarette that fell into the stuffing between sections of the sofa.

This case illustrates that one should not be too hasty in making a probable cause determination. Inside the first-floor closet, there was a second set of burn patterns on the walls, ceilings, and up to the bottom of the stair casing. In another locus there was a third burn pattern on the wood frame of a water bed. In the master bedroom closet, a fourth burn pattern was documented on the wood shelves used for clothes storage. In another bedroom a fifth burn pattern was observed on a wood dresser. There was no direct link between any of the five burn patterns. Diagnosis of exclusion ruled out electrical and mechanical sources of origin. It became obvious that this was a set fire. It is not known whether the arsonist put a cigarette between the sofa cushions to throw an investigator off track. Perseverance and thoroughness are vital to successful investigations.

Figure 3.6 Sofa springs, after the fire.

3.7.10 Computer Equipment Facility Fire (No. 20670)

A manufacturer of printed wire boards for computer equipment uses state-of-the-art high-technology equipment that transports the vertical process through tightly sealed wet process tanks by a carousel conveyor located above the process area. This technique allows finished boards to return to their original starting position. Thus a single operator, using a semiautomated load/unload station, attends an entire manufacturing system. (See Figures 3.7(a) and (b).)

Vertical processing achieves a high degree of product uniformity, because all panels are racked singly in an array that enables them to closely trail each other through the interconnected modular processes. Solution runoff is gravity enhanced and reduces dragout. This design significantly reduces ventilation requirements.

a

b

Figure 3.7 (a) Burned area of production system. (b) "Bread board" logic control system.

Transparent sealed side doors are provided on each tank for monitoring each process. At intervals, inspection stations are available. This decade-old technology has experienced several "generations" of improvements in design. The copper/tin–lead vertical processing system includes the following processes: load, dual develop, rinse, clean, rerinse, eight-step copper filtration process, rinse, tin–lead activate, dual tin–lead plate with final filtration process, rinse, strip, second strip filtration process, rinse, etch, etch (mixing process), final rinse, dry, and unload.

During a batch process that required the manual drain and cleaning of the sumps, a lead technician verified that the heaters were off. He cleaned the sumps and filled the tanks with deionized water. Sump 1 drain was left open. (There were two drains, one for each sump.) The tank was filled. The fill system was then turned off, and the technician left the plant for the day.

During the night, the tank drained. At 7 o'clock the next morning, the technician returned, but failed to verify the liquid level. He assumed it was filled as he had left it the previous evening. He turned on the heat. Fire broke out in a very short period of time. The fire started at the strip stations.

During the subsequent investigation, attention was focused on why the low-liquid-level switch failed to shut off the system. The switch was intended to maintain liquid level in the tanks by operating solenoid valves, to cut off the heaters to prevent fires, and to sound low- or high-level alarms to prevent tanks from overflowing. It was found that the plastic noncorrosive miniature level-limit control had a propensity for unreliability that had been tolerated by the management in the past. It was determined that, notwithstanding the actual tank level, the processor always indicated a "low"level.

In an attempt to debug the problem, the control switch had been replaced with two successive switch controls. Both had failed to register properly. A technical representative of the equipment manufacturer had been contacted. He had suggested a possible interim operating procedure to use the processor/software to "bypass all safety systems." As a result of this advice, the technician thought he was dealing with unreliable instrumentation, when in fact the real problem was an actual low-level condition. Lack of confidence in the low-limit control and the software bypass were the root problems.

This imprudent decision caused a massive fire only one week before a team of overseas customers was scheduled to visit the plant. In fact, an advance team was in route to the plant at the time of the fire. The expense of the fire damage was miniscule compared with the potential market loss resulting from erosion of confidence. Fire is a nonforgiving entity. The bypass of a safety system is akin to placing a penny behind a fuse, where the use of one cent can cause a tremendous dollar loss.

This investigation combined the techniques of fire investigation with the special knowledge and skills of electrical engineers specializing in computer software analysis.

Figure 3.8 Hand holding section of plastic downspout, showing sections of twisted heating cable inside downspout.

3.7.11 Do-It-Yourself Wiring (No. 20894)

The attractive townhouse condominiums were constructed for a New England community, in a latitude where winter freezing is the norm. Within a short period of time, the condominium builder/developer was plagued with an ice buildup at the flat roof and plastic downspouts. He sent his son to an electrical supply house, where he purchased several sets of heating cables. These cables were designed to melt the ice and allow for the smooth flow of water down the plastic pipes.

The young lad had no electrical training. He was not licensed as an electrician. He failed to notify the electrical inspector and did not obtain a building permit for the installation of the electrical heating appurtenances.

The thermostats were set at 42°F to provide a 10° margin of safety in avoiding freezing. They were fed from an attic light fixture outlet in violation of several NFPA-70 electrical code provisions.

During a site inspection after the fire, the heating cables were observed to be in a twisted mode (Figure 3.8). A test was conducted to replicate the scenario. The heat from twisted heating cables in proximity to the plastic pipes was the source of ignition. The lesson is simple: If you are going to "let George do it," make certain that he is a licensed electrician.

3.8 Conclusion

In the foregoing case studies, the clients who retained my firm for investigative work were not identified. The common denominator in all cases, however, was a client in trouble.

As a forensic fire investigator, the engineer first and foremost is a professional engineer. The client may be involved in a controversy with another party. The forensic engineer fulfills his or her mandate of independent commitment when the engineer identifies the facts forthrightly.

The path of a typical assignment identifies common procedures, whether the incident under investigation is a fire or a product liability analysis. Fire analysis requires an understanding of the glossary of terms unique to the fire service. An understanding of the lexicon of the insurance industry provides an enhanced ability to serve the clients, whose interests are inimical to the forensic engineer's professional tasks.

Clients who retain my firm are informed that it is the policy of the firm to share information with official agencies, including fire departments and the state police. A client who retains my firm to determine if he or she is innocent of arson should be aware of our policies. We share our information, good or bad.

The forensic investigator's work does not end at the fire scene. In point of fact, it just begins there. The term *working fire* is broad based. It includes the efforts of fire fighters knocking down the fire, the efforts of fire investigators on the scene, and the follow-through process. Persons must be interviewed. Evidence must be gathered and preserved. Documents must be reviewed and frequently rereviewed. There is a constant stream of information and data that must be digested.

Comprehensive fire investigations take time; and time is money. The client should be made aware of the cost that a proper investigation entails. Support of the fire department requires a large part of a municipal budget because communities recognize that it costs money to save lives and protect property. The forensic engineer and his or her clients should recognize that it takes time and money to conduct a complete fire investigation of substance.

Sometimes, at the end of an exhaustive analysis, it is simply impossible to determine the probable cause. The professional forensic engineer should not be embarrassed to acknowledge this. Speculation and possible causes have limited value in the quest for a reliable probable cause. Empirical experience, education, and professional development can improve the likelihood for success.

Information Sources

Eastman Kodak Company. *Basic Police Photography*. Rochester, NY: Eastman Kodak Co.
> Our technical library has the second edition, 1968. There are many other similar publications.

Kennedy, J. and P. M. Kennedy. *Fires and Explosions: Determining Cause and Origin*. Investigations Institute, 53 West Jackson Boulevard, Chicago, IL 60604.
> This is a 1505-page compendium that is a thesaurus and a resource of substance for every fire investigator. This reference is a must for one new to the field.

National Academy of Forensic Engineers. *Journal of the National Academy of Forensic Engineers* (NAFE, 174 Brady Avenue, Hawthorne, NY 10532).

All issues of this journal contain articles of interest. Milton Mickler wrote an article in Volume IV, No. 2, December 1987 titled "Forensic Engineering Analysis: Smoke Transport to Upper Floors During the 1980 MGM Grand Hotel Fire." (This writer worked on that case for another client and is pleased to commend Engineer Meckler for his work.)

National Fire Protection Association, Batterymarch Park, Quincy, MA 02269.

No one should be involved in fire investigation unless he or she is a member of NFPA and utilizes the volume of data produced by this organization. Of special interest are the *NFPA Fire Journal*, Codes, Standards, Reports, and Texts. The *Fire Protection Handbook*, and *Industrial Fire Hazards Handbook* should be part of an investigator's technical library.

Other Organizations

These organizations produce many publications and resources valuable to the fire investigator.

American National Standards Institute (ANSI), New York, NY
American Society of Safety Engineers (ASSE), Des Plaines, IL
Factory Mutual System (FM), Norwood, MA
Society of Fire Protection Engineers (SFPE), Quincy, MA
Underwriters Laboratory (UL), Chicago, IL

4. Industrial Accidents

RON HENDRY, P.E.

4.1 INTRODUCTION

In this chapter, "industrial accidents" refers to those mishaps that occur in the work environment and result in significant injury, product loss, or an adverse environmental impact. This covers an exceedingly broad range of topics. Injuries may range from those confined to one area of the body, to paraplegia, to fatality, or even to many fatalities. Also, the initial symptoms of an injury may be delayed and the injury may be progressively crippling. An example is the long-term effect of excessive exposure to asbestos and lead. The monetary loss from an industrial accident may range from thousands, to millions, and even to billions of dollars.

We are generally familiar with the large-scale mishaps involving the nuclear facilities at Three Mile Island, Pennsylvania, and Chernobyl, USSR, where harmful radiation was released. Most of us also remember the disaster at Bhopal, India, where a dangerous chemical escaped and killed over 2000 nearby residents. Only recently a large oil storage tank in Pennsylvania ruptured, allowing massive quantities of oil to contaminate a river that was a vital source of water for many people. These are large-scale situations where forensic investigators likely were hired to determine the causative factors of the accidents.

What usually makes much smaller headlines, and sometimes none at all, are the hundreds of thousands of injuries and product losses in the day-to-day work environment. Such things as the loss of a limb to rotating machinery or the loss of an eye to a flying piece of metal are devastating to the affected person. A small manufacturer can also be economically devastated if a storage vat fails and 2 weeks of irreplaceable product is lost into the sewer. On the other hand, a product manufacturer can be hurt and possibly put out of business if someone becomes injured using a product and eventually receives a large monetary award.

Many aspects of industrial forensic investigation are discussed in this chapter. The examples are drawn mostly from my investigation experience. Virtually all forensic disciplines can and do become involved in industrial forensic investigations. Thus, the window through which the reader of this chapter will view industrial accident investigations may unintentionally omit discussion of some important and interesting areas.

4.2   CATEGORIES OF INVESTIGATION

The following is a partial listing of the diverse areas of industrial accident investigation:

Building and construction activity. Some typical incidents revolve around accidents concerning cranes, scaffolds, aerial platforms, forklifts, elevators, worker lifts, and trench cave-ins.

Product damage resulting from equipment failures. An example is the loss of ice cream products when an ammonia refrigeration system goes out. Another example is the unexplained failure of expensive machinery, such as the collapse of an oilfield derrick.

Injury and/or guarding accidents. These may involve virtually any industrial machine, among which are presses, shears, and lathes. Also included are extrusion machines, blowmolders, automatic assembly process equipment, product transfer equipment, worker platforms, and worker lifts.

Electrical injuries. These involve accidents in which workers suffer burns and electrocutions from contact with electricity.

Fire losses. These include the investigation of fires associated with processes such as industrial ovens for food products. Also included are fires of arson origin.

Damage claims resulting from extremes of weather. These claims may be from flood or water damage and/or extreme heat or freeze damage.

Poisoning claims. These involve the exposure of employees to noxious or toxic chemicals. Also included are asphyxiations caused by the absence of oxygen.

Unexplained product loss/damage claims. A typical example is the unexplained loss of coal volume in slurry transit. Another example would be evaluating damage to expensive machinery resulting from exposure to impact loading during transit.

Environmental hazard situations. This might involve an assessment of water supply contamination caused by an oil spillage.

4.3 DIVERSITY OF TECHNICAL QUALIFICATIONS FOR INDUSTRIAL ACCIDENT INVESTIGATORS

Virtually all disciplines of engineering and the sciences may at times be involved in industrial accident investigations. A noninclusive listing is mechanical engineering, civil or structural engineering, electrical engineering, chemical engineering, safety engineering, industrial engineering, nuclear engineering, environmental engineering, human factors engineering, metallurgy, chemistry, toxicology, and fire cause/origin specialists. A brief discussion of several of the primary disciplines involved with forensic accident investigations follows.

Perhaps the widest range of investigations are conducted by those with a mechanical engineering background. This is partially because machines that cause losses are often mechanical in nature. The mechanical engineer is usually retained to investigate accidents involving machines or products that have moving parts or convert one form of energy to another.

Civil/structural engineers are usually associated with static structural failures such as structural or foundation failures of buildings.

Electrical engineers are associated with cases involving suspected electrical malfunctions. Typically the suspected malfunction may have caused a fire or an injury.

The metallurgist is usually involved when there is a significant question about the mode of failure of a certain component that cannot be resolved by visual examination. The metallurgist is also a good source for determining a part's material composition and its microstructure. The metallurgist often works in concert with a site investigator charged with the overall evaluation of the accident.

The chemist most often works in the laboratory. One prime activity is determining if accelerants are present in fire residue samples brought in from field investigators. This investigator also will venture into the field and take samples for lab testing. Typical situations involve questions of contamination of fluids, toxic substances in air, and failure analyses of engine oil.

The fire origin investigator can be an engineer or someone with a strong fire department investigation background. The usual responsibility is to determine the most likely origin of the fire from the remains of the burned building, structure, or machine. If a product is involved with the origin of the fire, then a mechanical or electrical engineer will usually examine the product to try to ascertain its failure mode.

When listing the disciplines involved in industrial accident investigations, we are making a list primarily of educational backgrounds. Of equal or even more importance is the experience background of the individual investigator. Elaboration on the experience necessary to conduct investigations is beyond the scope of this chapter. However, an investigator's education and experience are both open to challenge by the adverse attorney in discussing the investigator's capability to conduct an investigation.

There are areas of overlap between the basic disciplines. There are types of electrical product mishaps that the mechanical engineer can appropriately investigate. There are structural mishaps to mechanical machines that the structural engineer can easily investigate. Conversely some types of static structure failures can be analyzed by a mechanical engineer rather than by a structural engineer.

There are some fires associated with products that a mechanical engineer can competently investigate if the mechanical engineer has basic knowledge about fire investigations. The metallurgist and the engineering professional can at times overlap in the visual examination of failed components. The investigator with adequate knowledge can make human factor evaluations even though he or she may not hold a degree in that field. The investigator with some background research may render judgments about the adequacy of warnings or instructions, although this may not be a primary component of the investigator's background experience.

4.4 INVOLVED PARTIES

Industrial mishaps often have an effect on several parties in addition to those suffering the direct injury or loss. Other affected parties include anyone who may be financially responsible (those with "loss exposure"), as well as the plant owners, the employee union, governmental regulatory agencies, and even the public.

Plant owners are concerned with correcting any procedure or situation that may have contributed to the accident. They also need to assess the conduct of the employees involved for possible disciplinary action. If self-insured, then the plant owners may be concerned with costs associated with treatment and rehabilitation for the injured. If a fatality occurs, plant owners may be concerned about claims of gross negligence. Finally, they may be concerned about the possibility of government sanctions for safety violations.

The injured party desires compensation to enable satisfactory treatment and rehabilitation. The injured may have concern that his or her actions leading to the accident may be cause for some plant disciplinary action. Also, if a product can be found defective, then the injured party may be interested in compensation from the product manufacturer.

The original equipment manufacturer (OEM) of the product related to the injury is interested in the defense of the product against claims of defectiveness. The OEM also wants to evaluate any change to the product deemed necessary as a result of the accident.

The workers' compensation insurance carrier usually has the responsibility to provide the funds necessary for treatment and rehabilitation of the injured. This carrier often wants to know if an OEM should share or shoulder this expense because of a defective product. Sometimes another concern of this carrier is whether lax practices of the plant or the injured employee contributed to the likelihood of accidents.

The employee's union is concerned with worker safety and correction of the conditions leading to the mishap. The union also desires to protect the rights of the involved employees against unwarranted disciplinary action as a result of the accident.

The Occupational Safety and Health Administration (OSHA) is the primary governmental entity responsible for ensuring adequate employee workplace safety. OSHA is concerned with employee safety and whether safety violations may have occurred.

In situations where the machine involved in the loss was leased or rented, then the machine's owner has some concern for possible loss exposure. If the entity that earlier sold the machine is other than the OEM, then this firm also may have some concern about loss exposure.

In situations where the environment was contaminated by the release of toxic substances, then the public can have a strong interest in the resolution of the accident. This concern may involve questions as to who or what caused the accident and what measures can and should be employed to reduce the damage. In addition the public may be interested in developing appropriate strategies to minimize the potential for a recurrence of the accident.

4.5 GOALS OF THE INVESTIGATION

The goal of the investigation depends on the circumstances of the accident and what the client has requested. For example, I was once asked to restrict my examination of an aerial platform accident to the question of whether a malfunction of the machine contributed to the accident. Questions involving the actions of the worker and the accident layout geometry were answered by other investigators.

The most commonly desired information is the cause or causes of the accident. This includes any contribution to the accident by those suffering the injury or loss. It also includes any contribution by others, which often means determining whether or not a machine has any deficiency that may have contributed to the accident. Another common question is whether inadequate maintenance was a contributing factor.

Sometimes the investigator is asked to compare the statements made by witnesses and injured parties with the story told by the physical evidence. On some occasions the investigator may be asked if an illegal or improper activity was causative. On other occasions, an investigator may be asked to assess damage to the products or material. Another possible request is for an evaluation of a repair estimate for the damaged products. In environmental contamination cases, the investigator may be asked to assess the negative impact on the environment.

4.6 ETHICAL CONSIDERATIONS

4.6.1 Before Taking the Assignment

Several affected parties may be interested in an independent forensic investigator. The investigator should be aware of the interests of a potential client when he or she discusses an assignment. It is important to understand why the client wants an investigation.

The investigator has certain ethical considerations that should be addressed before accepting any assignment:

Staying within boundaries of reasonable competency. An investigator should possess an adequate background of education and experience to be able to handle the assignment. This often includes performing the investigation, consulting with the client, and, if needed, giving expert testimony at some later time.

Capability to perform an objective analysis. All of us have some bias. Objective investigators realize their natural biases and work to neutralize them for the purposes of an investigation.

Capability to make difficult decisions. The investigator must have the fortitude to formulate sound opinions and the ability to defend those opinions against attacks from all sides. At times, these opinions may be contrary to what the client would like to hear.

Capability to render a timely opinion. Often a timely response can mean considerable savings to the client. This often is the case for clients who are found to have loss exposure. Early findings of loss exposure can lead to a favorable negotiation that limits the total monetary outlay.

Knowledge of legal processes. This knowledge helps the investigator focus on the relevant areas of the investigation. An informed investigator is better able to cover the necessary bases during an investigation. Often the investigator can obtain the minimum relevant legal information from the client when taking the assignment.

4.6.2 Wearing More Than One Hat

The investigator who agrees to provide an investigation for a client usually wears at least two hats. The first hat is that of a consultant to the client as to the things important to the client's best interests. The second hat is that of an unbiased expert investigator who will develop unimpeachable opinions.

When wearing either hat, the investigator has ethical obligations to other parties. The investigator's opinions as to who is at fault can result in an expense to the defendant parties. The expert's opinions about how an accident happened can also adversely impact an employee's standing with the employing company. Thus, the investigator should have strong bases for any statement regarding contribution to an accident.

4.6.3 The Investigator Must Have Bases for an Opinion

For expert investigators, an opinion without bases is not worth giving. An attorney once came to me with an injury accident problem. He was defending the manufacturer of an industrial pushcart. A woman using this cart had reportedly injured her back when a wheel unexpectedly came off. The attorney wanted my evaluation of the cart and whether I would be able to testify in its defense. The trial was to take place in the very near future.

My investigation and tests found the cart design to be clearly defective and likely a contributory factor to the accident. The attorney was thus informed and my efforts were terminated.

A few weeks later, this attorney telephoned in an obviously good mood. The plaintiff had refused his modest settlement offer. However, he had gone ahead with the trial because the plaintiff's expert had not done his homework. This expert had said the cart was defective, but could not provide any substance or basis for making that statement.

Thus, this attorney won the day for his client because he knew the other side had not done their job and could not provide any basis for their claim. This example also illustrates that the client very often benefits from solid investigative work, even if the investigator finds the client at fault.

4.6.4 Referring the Client to the Type of Expert Needed

Investigators routinely assist clients in determining the proper expert for the job. Recently, I was consulted about an accident involving an earth-moving machine that fell into a lake. The reported story was that the machine was required to push lignite ash into the lake from a landfill built up from ash material. I advised the client that he should find an expert on soil stability. This client later related to me that a very successful conclusion to that litigation was based on the findings of the soil stability expert.

4.7   INVESTIGATIVE TECHNIQUES

There is a simple logical order to most investigations. The major phases can be identified:

Information gathering

Site or product examination

Analysis and testing

Opinion formulation

Giving the opinion to the client

Some of the phases overlap or occur at the same time as others. This is especially true of the information-gathering phase. The following discussion outlines the typical procedure for many injury accident investigations.

The investigator acquires initial information about the accident from the client. This usually includes a verbal description of the accident and a transfer of all pertinent documents, photographs, and witness statements.

The investigation phase is usually begun while the initial information is gathered. Preferably, this includes an on-site examination of the location and equipment involved in the accident. At the accident site, the investigator obtains plant employee opinions about the circumstances of the accident. Pertinent operator and maintenance manuals concerning the machine involved are also reviewed. If possible, the investigator should obtain copies of the pertinent portions of the operator and maintenance manuals for later review.

A responsible plant employee usually shows the machine and/or location of the accident to the investigator. The investigator should inquire as to any accident scene changes that that may have been made after the accident and before the examination. The investigator should keep a list of all parties consulted, and roughly what they have said.

The preferable situation is for the investigator to monitor the machine in operation in a manner similar to that occurring at the time of the accident. Of course, any operation should be only in the best interests of safety for everyone present at the inspection.

The investigator should document the accident scene and the pertinent equipment. One method is by good-quality photography using a 35-mm camera. Recently, camcorders are finding increased usage in documentation of accident situations. One important documentation step is to obtain and write down all pertinent machine identification information. Also, the investigator should measure and write down all the relevant measurements that may later be needed for layout analysis and/or evaluation.

A clear statement or understanding should be obtained from plant personnel as to the responsibilities and duties of the employees involved in the accident. This statement should also include what the employees were supposed to be doing at the time of the accident.

As part of the information gathering, the relevant industry standards for the subject machine should be obtained and reviewed. As a rule, industry standards are viewed as the minimum acceptable standards for manufacture of that class of product. As such, they are not generally regarded as the final statement on the acceptability of the safety of a product. However, these standards often provide useful background information.

Any applicable OSHA standards for the subject machine should also be reviewed. Often the investigator finds that the OSHA standards closely correspond to the industry standards. In some cases, such as fatality accidents, an OSHA field investigator will investigate the accident and file a report. This report can be obtained by filing a request under the Freedom of Information Act.

The next phase is that of testing and/or analyzing the data. Testing may require an additional trip to the plant to conduct a test not considered in the first visit. Usually, in cases involving guarding situations, little testing is done.

Analysis involves a review of all the information obtained. It sometimes includes consideration of human factors for the work environment. Such considerations as visibility and placement of controls may be important. Where a failure has occurred, a process of failure analysis is used to find the contributing causes. The techniques of failure analysis are discussed in more detail in Section 4.9.

The next phase is to form an opinion as to contributing causes. Several questions must be addressed. These vary depending on the case, but may include the following:

Who or what contributed to the causation of the accident?

What part did the injured play in the accident?

Was the machine defective in any manner?

Was that defective condition a cause of the accident?

Did poor maintenance contribute to the accident?

Was the maintenance manual adequate?

Did the machine have adequate warnings?

The last activity phase is giving the firm opinion to the client. In many cases, the client should be contacted immediately after the field examination. The client usually wants to know if the investigator feels that continued work might prove useful to the client's interests. In some situations the experienced investigator is able to recommend a cessation of work immediately after the initial examination. This should be done only if the investigator is very sure of his or her finding. The investigator should always refrain from giving snap judgments about an accident until enough facts are available to form a firm opinion and he or she is comfortable about the opinion.

It is extremely important to first give a verbal communication of the opinion to the client. In the case where the expert's opinion is adverse to the client's hopes or posturing, the client may not want a written report. The client may want the expert to send the report to legal counsel. Alternately, some clients do not want a written report for any reason. Preparation of a written report is discussed in Chapter 11.

4.8 INVESTIGATIVE TOOLS

Many tools are available to the investigator. These range from the simple to the exotic. A partial listing of a few of the more common tools follows:

Measuring devices of any and all types including tapes, transits, levels, stopwatches, and hardness testers

Sampling devices of all types including those for fire residue, air, and soil

Photography equipment (the most common item is a good-quality 35-mm camera with a variety of lenses; recently camcorders are finding increasing usage)

Low-power microscope

Scanning electron microscope

Element and chemical analysis devices

Fire residue analysis equipment

Computers

Drafting table and equipment

4.9 INVESTIGATIVE ACTIVITIES

4.9.1 Failure Analysis

Failure analysis is the process of determining the order and timing of important events leading up to and through the primary failure event. The investigator looks at such things as the distortion or lack of distortion of parts. Also checked are the failure modes of individual pieces or components including types of failure and failure direction. In addition, the investigator checks for the presence or absence of corrosion on failed surfaces, scratch or damage markings and their movement directions, and whether any parts were missing.

In performing this analysis, the investigator usually works backward from the time of the accident. First, the activities and changes to the accident scene or product failure from the time of the accident to inspection are reviewed. Many issues can be uncovered in this review. For instance, a fallen structural beam may have been cut into pieces by a torch and relocated to help rescue an injured worker.

Next, the investigator usually determines the sequence of events that transpired during the accident phase. If a crane boom collapsed, questions will be investigated relating to the initial point of collapse, activities that took place during the collapse, the adequacy of the foundation for the crane, and so on.

The final phase of the analysis is the evaluation of preexisting conditions and determination of the causes promoting the failure. What activity set the accident into motion? Was there an ongoing structural failure in the boom that inordinately weakened it? Was a vital structural component missing? Was the foundation setup inadequate and already giving way? Had some vital safety consideration or procedure been ignored?

4.9.2   A Failure Analysis Example

Close visual examinations of suspected malfunctioning parts often yield a rich story of the events leading to the accident. For this reason, it is extremely important to avoid disturbing the accident site or suspected device.

One time I was assigned to examine a chain fall hoist on which the chain separated and caused a fatality accident at a construction site. The chain fall hoist had two chains: one that did the work of lifting and lowering, and a smaller one used by the worker to operate the hoist. The hoist lift chain held a platform used for transferring material from one floor to another. An employee ventured out onto the platform to get the small chain. The hoist lift chain apparently separated while he was on the platform. The man fell to his death. The actual chain link that separated was never recovered.

Loss of the involved link was not critical to the investigation. Field personnel had done a good job of not disturbing the evidence on the hoist after the accident. My later examination found numerous marks on the hoist that told a descriptive story of why the hoist lift chain failed. These marks revealed that the hoist lift chain had been allowed to double upon itself and someone had tried to pull it through the hoist in that manner. This caused one of the chain links to open and separate. This chain was much weaker when allowed to double upon itself. It was not the type of chain one would recommend for the hoist.

4.9.3   Failure Analysis Using Photographs

Sometimes the investigator is brought into an assignment a considerable time after the accident has occurred. In those situations, often all that is available for early analysis are photos of varying quality. These photos may show failed parts that have since been destroyed by repairs to the machine. They may also show a toppled machine with hopefully enough detail to eventually enable the investigator to determine the cause of its overturning.

As an example, truck-mounted boring rigs are used in the oilfield to make large-diameter starter holes at oilwell drilling sites. Once an operator of one of these devices was injured when parts of the mobile boring rig came apart during an operation. I was given the investigation assignment on behalf of the manufacturer of the rig. I was supplied with some field photographs that were taken at the accident scene before anything was moved.

Later, a trip was made to the storage yard to examine the damaged pieces of the rig. Connection failures were found in several parts. Several failures were progressive in appearance. A study of the field photographs revealed that an important stabilizing component was missing at the time of the accident. This item was an auxiliary hydraulic cylinder that had been removed for repairs. The final resting position of the separated parts in these photographs also suggested the likely sequence of separation. Eventually, the sequence of events leading to the accident was determined. Unfortunately, I found several deficiencies with my client's design of the equipment.

4.9.4   Tests and Simulations

For some accident investigations, the ideal would be to operate the subject machine as it was operated at the time of the accident. This can provide good insight into the problem. If this is not possible, a similar machine is sometimes operated with beneficial results. However, simulation of an operation can be impractical for logistic and/or safety reasons.

A structural or performance test can be made of a similar material or component that may have failed. This is often done in laboratories that specialize in such tests. These tests can help in the evaluation of the adequacy of strength of some suspected component.

4.9.5   Layout and Analytical Analysis

In certain stability problems such as crane tip-overs, some analysis may be required to determine the cause of instability. Where a structural item has failed, a rigorous analysis will usually be made to ascertain the adequacy of the strength of the part.

A scale layout can resolve problems in understanding the events of an accident. I recently used such a layout to good advantage in reconstructing how and why an aerial platform rolled into a drainage ditch. Oblique elevation drawings and perspective drawings were invaluable in supplementing the plan view of this accident scene to illustrate and prove the sequence of events. The adverse expert relied on mathematical computations for his analysis. In addition to slight inaccuracies in his computations, he had a decided disadvantage in the visualization of the accident.

4.9.6   Use of Multiple Investigators

Many investigations require more than one investigator to handle the various phases. The investigation may require the services of a metallurgist to investigate the material adequacy or failure mode of a certain failed component. The fire specialist may need the services of a chemist to analyze fire residue for the presence of accelerants used by suspected arsonists. On other occasions a mechanical engineer and electrical engineer may both need to examine a complex electromechanical machine.

The metallurgist often performs an invaluable service related to an accident investigation. A competent engineering investigator can overlap with the metallurgist on failure analyses down to the low-power microscope level. Even at this level, it is often wise to have a metallurgist visually examine the failure. The metallurgist usually should have prime responsibility for examinations beyond the level of the low-power microscope. For instance, the metallurgist may use the scanning electron microscope to do a fractography analysis of the failed surface. This analysis uncovers many definitive failure features that a low-power microscope cannot discern. Among other equipment of the metallurgist are devices that enable the determination of the basic material elements and the chemical composition of the failed part.

On occasion, a forensic chemist can provide invaluable assistance to the engineer investigator. Perhaps the most notable use is in determining if accelerants exist in fire residue samples. However, chemists can also analyze virtually any fluid for indications of contamination or indications of why something such as an engine failed. They can assist in corrosion problems. Chemists can perform chemical analyses on applied paints or protective coatings to resolve disputes about their adequacy. They can perform air or gas analyses to check for toxic levels of chemicals, such as formaldehyde.

4.10   INDUSTRIAL ACCIDENTS INVOLVING KNOWLEDGE, ACCOUNTABILITY, AND JOB DESCRIPTION

In some injury accidents, the injured party's awareness of danger associated with a machine and/or his or her job activity at the time of the injury are important factors. Following are some observations regarding these factors.

4.10.1   User Knowledge and Accountability

Accountability for actions should increase as the operator's knowledge of a machine's danger increases. The experienced worker should possess considerable knowledge about the machines the worker operates and should have considerable accountability for his or her actions. Accountability for such a worker should ordinarily be greater than that for the inexperienced worker or for the typical ordinary citizen consumer using the same machine.

I used this argument in a case involving a wood router to successfully defend the machine's lack of printed warnings. The intent of warnings should be to inform the user about a danger he or she may not otherwise suspect. The plaintiff was an experienced wood shop employee who had been injured by a splinter of wood thrown off by the wood router. My opinion was that any experienced wood-working shop employee should know that wood routers have the capacity to eject slivers of wood that can cause injury.

The experienced worker who continues to use equipment known to be damaged, knowing that an injury is possible, should be held responsible for assuming risks. The owner of a sewer cleaning firm once received a permanent hand injury when an auger cable broke in his hand and literally screwed itself into the hand. My examination of the cable found that it had several preexisting visible kinks and that the failure occurred at one of the kinks. The owner was in fact holding the cable at the kinked location to steady its rotation. The owner later testified that he knew that other people had been injured when their sewer cables broke. However, in this case, the owner eventually received a substantial settlement, as the result of several other contributing factors.

4.10.2   Differences between the Maintenance and Operator Functions

Generally, the primary protection for someone performing maintenance or cleaning is procedural. The reason for this is that the maintenance function often requires the worker to come into close contact with moving or electrical parts of a machine. In contrast, the operator's work station generally should be designed or arranged so that the operator does not come in contact with the zone of operation of the machine. Ideally the operator should not be exposed to an accidental mishap involving a machine. Such is not normally the case for maintenance operations.

For example, an employee was once engaged in cleaning a large food mixing vat. This vat commonly mixed a corn product. Deep inside the vat was a rotating metal blade assembly. On this occasion, the employee physically climbed into the vat and was cleaning the blades. Prior to getting into the vat he turned off the wall switch control. This wall switch held the operating controls for several food mixers. While the first employee was working away in the vat, another employee came into the room and activated the corn mixer wall switch by mistake. The first employee's legs became trapped in the rotating blades and he was severely injured. This injury would have been prevented had the wall switch lockbox been locked to prevent accidental activation.

4.11 GENERAL OBSERVATIONS: HUMAN NATURE AND ACCIDENTS

People are not perfect. We sometimes fail to do what is best for us. Sometimes this is because we do not see the relevance of decisions affecting safety. Other times it is because we do not think anything can happen to us. The following observations are based on experiences with a number of injury investigations.

4.11.1 Safety First?

Human factor experts tell us that people generally do not proceed on the basis of "safety first" in planning their actions. Instead, they generally operate on the basis of an assumed low probability of injury for the action desired. The problem with this is that an operator may perform an excessively dangerous procedure a thousand times without injury. Unfortunately, it only takes one accident to lose an eye, a hand, or a life.

I once examined a small sliver of metal that cost a worker his eye. The metal had flown into his eye as a consequence of using a hammer to strike a large metal casting. It is difficult to estimate how many times one would have to do this operation before a flying metal sliver came off and struck something as small as an eye. However, we all know that this type of accident is easily avoided if a worker simply takes the time to put on safety goggles.

4.11.2 Procedural Errors

Many of the examples discussed in this chapter contain the element of failure to follow a safe procedure. This cannot be emphasized too strongly. In many accidents an investigator may find some product deficiency that was contributory. Yet, in those same accidents, there is often the omission of a safety procedure that would have prevented the accident.

Once a worker on a forklift moved out onto a parked flatbed trailer to retrieve some bailed paper products. The trailer had been backed up to the dock and uncoupled from the truck tractor. Two landing gear jacks were supporting the trailer on the uncoupled end. As the forklift reached the region of one of the jacks, the jack suddenly collapsed, allowing the trailer bed to tilt. The forklift toppled over and the worker was pinned and crushed to death by the forklift rollover bar.

There was a design question about the landing gear jack; however, had the worker followed the recommended industry procedure of placing external jackstands under the trailer, the accident could have been avoided.

4.11.3   Miscalculations

Snagging is one of the most frequent miscalculations of employees. Loose clothing and rings are the primary offenders. Employees often fail to appreciate the danger of certain actions until it is too late. I have worked on several cases involving rings that become snagged, resulting in the loss of a finger or hand. One young lady lost her left hand to a paper shredder when her engagement ring became snagged on exposed feed rollers. Another lost her ring finger when her ring became snared on the top edge of a shelving brace as she jumped to the floor from her perch.

Years ago, a welder in a large shop needed to do some drilling on a part he was welding. So he took the part over to an unoccupied large drill press and proceeded to drill a hole. Welders typically wear loose clothing and gloves to prevent being burned from weld spatter. This welder did not take the time to take off his welding gloves. The drill snagged one of the gloves and proceeded to rip his arm off.

It is an unfortunate commentary about workplace injuries that accidents such as these are destined to be repeated over and over again. It is hard for many of us to realize the importance of following safe practices. The fortunate ones are those who have close calls and learn from experience the value of following safe procedures.

4.11.4   Fewer Accidents Happen When All Runs Smoothly

When everything is running smoothly, accidents usually do not occur. The raw material feeds into the machine smoothly as intended. The machine makes the product correctly. The conveyor belt transfers the product as intended. The packaging equipment performs flawlessly. Normal functioning of equipment does not require the operator to be concerned about correcting a problem.

When something goes wrong, however, a situation is created where the operator may try to interject his or her hand or body into the point of operation zone and risk injury. An example of this occurred on an aluminum extrusion machine. The operator had the task of applying some lubricant to the nose end of the hot billet (heated block of aluminum) while it was on a conveyor on the way to the extrusion machine. On one occasion he failed to do this at the proper time. So he grabbed some lubricant and reached over to apply it, just before the billet reached the extrusion die. His glove became snagged on the billet preventing him from retracting his hand. The extrusion machine severed his hand. I had the task of determining if the opening that allowed the worker to reach the area just in front of the extrusion die constituted a guarding deficiency. My opinion was that the opening was guardable without adversely affecting production. Thus, the extrusion machine was found to have a guarding deficiency.

4.11.5 Using the Machine Even Though It Is Not Running Normally

The continued usage of a machine showing malfunction symptoms is unfortunately a rather commonplace occurrence in injury accident investigations. There may be several reasons. One may be the lack of machine function knowledge on the part of the opertor and even on the part of maintenance personnel. A man who lost his hand to a liver organ press later testified that the machine had malfunctioned before his accident. A woman whose hand was severely burned by a laundry ironer made the statement that it was making strange noises and was acting irregular prior to the time her hand became trapped.

On another occasion, an employee was operating a blowmolding machine with which he was not very familiar. He was experiencing difficulty in getting the process to start. He had earlier found out from an another employee that if he put his hand over an exhaust port on the die, the product would stay in the correct position. Thus, this practice was sometimes carried out. On this particular occasion, the dies had been changed and were narrower when closed. When the operator placed his hands on the exhaust port and the dies came together, his hands became pinched by the surrounding parts. He suffered a severe hand injury.

The question in this case was whether the opening that allowed the operator to get his hands into the machine was a guarding deficiency. To reach this opening, the operator had to stand on a bucket and lean over the conventional front guard. In some plants, however, a worker's platform could have been constructed that would have been the same height as the bucket. That factor became part of the basis for my decision that there was a guarding deficiency.

4.11.6 Accident-Prone Workers

Some workers are more accident prone than others. This can be attributed to use of poor procedures. It can also be a function of the type of work they do. There is an old deposition summary of one worker that has been passed around. This summary outlines the injury history of a worker who had filed suit against a hand-held ripsaw manufacturer because his knee had been permanently injured in an accident. His contention was that the guard had jammed open and allowed his injury to occur.

This worker detailed an exceptionally long history of injuries. Many were associated with the type of saw that caused his injuries. One can criticize his poor procedure in the use of the saw. However, when we think of the countless times he had used the saw in his work, we should expect that some injuries would occur just as a function of his high-risk activity. Thus, some accidents can also be looked upon as a natural by-product of action.

4.12 ORIGINAL EQUIPMENT MANUFACTURER RESPONSIBILITIES AND PROBLEMS

When a significant loss occurs in association with a machine, this machine is often examined for some contributing defective condition. The OEM can be faced with claims ranging from the "nuisance claim" to the valid. After evaluating many such machines, I have developed some general observations regarding OEM problems and responsibilities. Several of these observations are discussed in the following sections.

4.12.1 Designing for the Emergency

Designing for emergency is an area where some manufacturers have room for improvement. On a recent case, the OEM defended the radical separation of the worker's basket of its aerial platform by saying the machine was abused beyond the normally expected usage. The manufacturer should expect some misuse of such a product. The line between foreseeable misuse and unreasonable abuse is not easy to draw.

In this case, the only major structural damage was to the parts connecting the worker's basket to the boom. The analysis by the defense expert weakly supported the manufacturer's argument. That expert's opinion revolved around marginal computations.

My contention was that the basket attachment design was defective by definition and this was supported by cursory inspection. The design lacked ruggedness. The basket and boom sustained little or minor damage in the accident. If a manufacturer is producing an aerial platform that will enable workers to safely work and maneuver, then the manufacturer should not allow the basket to completely separate under reasonably foreseen emergency conditions. The basket and boom should experience significant deformation damage before radical separation.

4.12.2 Unevaluated Product Changes

This is another frequent problem area for OEMs. A modification to existing equipment can potentially render a good product defective. This happened to one OEM when the company produced a drill adapter as a customer service. The problem was that this adapter enabled the customer to overload the basic OEM equipment.

In another case, an OEM desired to move into a new market and thus made a small change in a material specification; however, this small change rendered its product defective.

4.12.3 Who Provides the Guarding?

The OEM is usually looked upon as the most able and capable of providing any necessary safety guarding features. Most often, guarding should be provided by the OEM, but there are some exceptions.

There are some situations in which the OEM may be justified for not providing any and all types of guarding. One case could involve options between two or more safety features, where only one safety feature is mandatory. Years ago, it was an accepted practice in some areas to not make any optional guarding features mandatory. Nowadays, it is difficult to justify making safety equipment optional.

Another case would be where the manufacturer is supplying an unfinished machine and does not have the end-use information necessary to provide the safety feature. Here there is a point of view that the manufacturer of the basic machine still has some duty to participate in the process of ensuring the operational safety of the machine.

4.12.4 The Open and Obvious Hazard Argument

One OEM defensive argument is that the hazard was open and obvious. This is a debatable defense. For one reason, some hazards pose very subtle dangers, even though open. The manner in which a person becomes snagged or trapped is often not obvious or anticipated. One key question is the degree of difficulty for the OEM to have provided guarding. If guarding is not feasible or practical, then the open and obvious argument may have merit, even against the lack of warnings.

4.12.5 Changes in the Intended Mode of Operation

A situation sometimes arises where a machine has been rendered dangerous because someone changed its intended operation mode. One example is a wastepaper box bailer designed to lie horizontally. This box bailer was purchased from the original owner, transported to another location, and installed vertically. This created a most undesirable exposed hazard that eventually severed the leg of an employee who slipped.

Another example is a machine composed of several large components that interface together into an automatic operation. At some later time, a basic machine component may have been either sold or moved and set up without the other components. A highly exposed point of operation hazard may now exist as a result.

4.12.6 Mismatch between Provided Maintenance and Needs of the Machine

Often, trouble with a machine's operation results when the level of capacity of the operator and maintenance personnel is not adequate for the needs of the machine. To the extent possible, the manufacturer should take into account the competence level of those operating and maintaining the equipment. That level of competence can, unfortunately, be minimal.

One such example was that of a liver press. This was a machine that compressed animal livers into a salable shape. The operator of this machine had his hand severed one day by the press ram. He had reached his hand into the operation zone as usual to help push the product out. This was a very undesirable practice. This time, the ram became activated and severed the operator's hand.

On examination, the machine was found to have been poorly maintained. A safety hood had been designed by the manufacturer and offered for this machine as an additional item years after the machine had been purchased. This safety hood had already been purchased and received, but it had unfortunately not been installed before the accident.

The installation of the hood was clearly beyond the capacity of the regular maintenance personnel. It was actually beyond the capacity of the maintenance personnel, in this case, to adequately maintain the regular machine.

4.12.7 Feedback

Sometimes lack of feedback about a deficiency prevents the OEM from discovering a major design problem until a serious accident occurs. Purchasers of equipment often make their own repairs rather than resort to complaining to the manufacturer. They do this to get the machine back into service faster. A related problem occurs when the user repairs are inferior to the OEM parts.

4.12.8 Machines That Don't Wear Out

Many accidents happen on older machines that may have been sold several times. These machines have often lasted longer than one should expect. Also, these machines may have undergone major modifications. Nonetheless, accidents do occur on these machines that can be attributed to deficiencies in OEM design or manufacture. A philosophical question exists as to whether such findings place an undue burden on the manufacturer.

I investigated an accident involving an old upright press. The setup operator had just been injured when a major casting member had split into two pieces. This casting was part of the primary stroke force assembly. It was original equipment on the 40-plus-year-old machine.

The setup operator had been trying to index the press to make a cup type stamping. He had trouble getting the stamped cup to separate from the original surrounding material. He misinterpreted the problem as requiring more stroke from the press. Had he been more experienced he would or should have seen that the machine stroke was already bottoming out on the stamping die. The sample parts made just before the accident clearly reflected that situation.

He increased the stroke again. On the very next attempt to make a part the large casting split in two and fell on the operator. Later, it was determined that the actual problem preventing the cup separation was internal within the stamping die.

I was given the overall assignment of evaluating the casting failure. Several large voids and unusual casting features were found. A metallurgist also examined the casting. He also found an excessive amount of casting voids and irregularities. It could also be seen that one large void opening had been patched and plugged to make the casting salable.

Some voids in large castings of this type are to be expected. The question was "Were these voids excessive to the point of being defective and contributory to the accident?" My opinion was that there would be no question had the accident happened shortly after the machine had been built. Another question was whether a large part should be allowed to fall on an operator in such a mishap. I eventually came to the conclusion and later testified that the casting was defectively manufactured, although it had successfully performed for a long period of time. Obviously, the operator's actions also contributed to the accident.

4.12.9 Not All Accidents Are Avoidable by the OEM's Efforts

One should not assume that all accidents to operators of equipment are avoidable through the OEM's efforts. An example is the case of a wood router accident I investigated several years ago.

A simple wood router of considerable age had been set up as a shaper to shape the profile of long strips of wood. The basic machine was nothing more than a rotating shaft protruding vertically upward beyond a horizontal bench base. The shop owner had made his own jig device to hold the work piece in place while the shaping was being done. The shaping blade knives were mounted onto the rotating shaft.

One day a shop worker was manually feeding the wood strips into the router. Suddenly a long thin sliver of wood came flying back out of the entrance chute. This sliver imbedded itself in the worker's hip. A later infection resulted in a frozen hip joint.

In my examination on behalf of the insurance carrier for the wood router, I was unable to fault the basic machine for this accident. My opinion was that the OEM could not have provided any practical safety device that would have provided reasonable assurance against the occurrence of this type of accident. The plaintiff produced a couple of experts in the ensuing trial that disagreed with my position. The defense prevailed, as the jury found no fault with the product.

As frequently is the case, it appeared to me that the wrong party was being sued. If the injured worker had a valid complaint, it likely was against the physician who treated him initially and failed to remove all the splinter material.

4.13 OTHER INVESTIGATIONS: CASE STUDIES

The following case histories show some of the variety encountered in industrial accident investigations.

4.13.1 How Much Visibility Is Enough?

The following case illustrates the wide variety of allegations that OEMs have to address. Once I was given the assignment of evaluating the design adequacy of a mobile crane cab on behalf of the insurance carrier for the manufacturer. The specific question was whether the visibility from the operator cab was deficient. A severe electrical burn injury had been suffered by a ground employee when the crane boom had contacted a power line. This employee unfortunately was in contact with the crane when it touched the high-voltage line.

The crane was being moved to another work site. The injured was one of several ground personnel walking beside the crane to the site. One of their duties was, or should have been, to provide assistance to the operator about such things as clearance under and around power lines.

The crane was several years old. Newer model cranes had improved cab visibility when compared with the older model. I was given excellent photos of the accident site as well as some important site dimensions. The cab manufacturer supplied me with the design dimensions for the cab. The crane was later examined and several visual tests were performed.

My evaluation determined that the operator could easily have seen the location where the crane tip contacted the power line. However, he likely would have been required to move forward some in his seat to see the line. His visual perspective would have been inferior to those on the ground. The ground personnel should have been giving the crane operator guidance and should have stayed clear of the crane. No contributory deficiency was found on the part of the crane.

4.13.2 Gross Negligence Actions

Workers' compensation insurance usually insulates the employer against civil suits in all situations except the claim of gross negligence in fatality accidents. Reportedly, it is not easy to get the jury to find against the employer in these actions.

I have investigated two such accidents on behalf of plaintiffs. One situation involved a maintenance helper who was struck by automatic concrete batch equipment while idly watching his fellow employee work. The other involved an employee taking inventory who was knocked into a foundry cupola by an accidentally activated fuel-loading bucket. Serious safety deficiencies were found in both situations. Also found in both cases was considerable contribution by the victims. One action was unsuccessful, and the resolution of the other is unknown.

4.13.3   The Painted Scaffold

Once I had the assignment of determining the cause of a scaffold accident. Some painters had been given the job of painting the exterior of a building. They had rented an exterior scaffold and installed it on the building. This scaffold used a cable-and-winch arrangement for raising and lowering the work platform. During the course of their painting, the workers and scaffold fell from the building. I was asked to determine the cause of the accident.

On examination of the scaffold shortly after the accident, a malfunction of the winch and cable was quickly dismissed. The falling paint had covered and preserved the winch and cable positions as the unit was falling. Both cables were extended the proper length for level operation. It was ultimately found that the workers had used sandbags to secure the cable support poles and that these sandbags had eventually worked loose. The painters had not secured the scaffold support poles to any substantial structural elements on the building.

4.13.4   Water, Water Everywhere

There is perhaps nothing less desirable to an owner of paper products than having tons of water cascading into a paper storage warehouse. Some time back this did happen. An overhead water pipe coupling failed and separated, allowing water under 150 psi pressure to pour into a paper storage warehouse for several hours. The location of the failure was obvious. One coupling casting had fractured in service. It was obvious that the fracture was progressive in nature. One could also easily tell which side failed first from the rubber impression on the coupling seal.

A site examination was conducted and the accident scene was documented. After making a visual and low-power microscope examination of the coupling, I gave it to a metallurgist for closer examination. His examination found voids that are commonly associated with disturbance of the casting before its solidification at the foundry.

No alteration or abuse to the coupling was detected. No installation irregularity was likewise discovered. This case was recently settled, as the defendants had little to argue over except for the amount of the compensation and its apportionment among them.

4.13.5   Ammonia-Flavored Ice Cream

The manufacturer of some popular ice cream began having some serious problems with his warehouse refrigeration system. This system used liquid ammonia evaporation. Three failures occurred in a short period of time. In two of the failures, the ice cream in the warehouse was contaminated such that it had to be discarded.

I examined the system shortly after the initial failure. A refrigeration repair service had already rewelded a failed liquid ammonia tube junction box. Soon after my examination, the rewelded box failed and another warehouse of ice cream had to be discarded. This time the tube junction box was totally removed for closer examination.

With the help of a metallurgist it was easy to establish evidence of poor welding as a promoting cause of the last failure. Additionally, there were indications that the initial OEM's welding of this rectangular-shaped tube junction box was substandard. Added to this was the OEM's utilization of a poor pressure vessel shape for the tube junction. In evaluation of the system, the absence of a relief valve was found to be a possible contributing factor to the failures. Thus, the investigation found both the OEM's refrigeration equipment and the weld maintenance to be deficient and contributory. The ice cream manufacturer's claims were swiftly settled once the report was published.

4.13.6   Anti-Two Block

Mobile cranes use a cable, winch, and hook arrangement to lift and relocate loads. Sometimes, above the hook on the cable end is a weight called a "headache ball" (for good reason). If some activity causes all slack between the cable hook and boom nose to be eliminated, a condition known as "two blocking" results. If allowed to continue, the hook and headache ball will be pulled off and fall, possibly injuring someone.

Some time back this happened to a mobile telescoping boom crane that was in the process of removing a rock crusher from a lowboy trailer. This impressive looking crane had a double-hook, double-winch assembly. Only one hook and winch was being used. The unused hook became "two blocked" as the boom was being extended. The situation went undetected by the ground signalman. The headache ball fell off the crane and injured the signalman.

This crane had two warning devices that should have helped prevent this accident. The only problem was that both were optional, and neither had been installed on the subject machine. Compounding the problem was the fact that the crane was rented and was being used by a relatively inexperienced crane operator and signalman.

Two blocking can occur on any crane. This hazard is something crane operators should always have in mind; however, a telescoping boom crane with a double hook can tax the ability of an inexperienced crew. My opinion was that a double-hook telescoping boom should have an anti-two block feature as standard equipment.

4.14 IMPACT OF FORENSIC ACTIVITY ON IMPROVED PRACTICES, PRODUCTS, OR PLANNING

I am unable to cite known specific examples where my forensic activities have advanced the cause of reducing the severity and magnitude of failures. Hopefully that has happened. Once an assignment is completed, there is often little feedback about end results. Beyond that, there is even less known about changes that may have been made as a result of an investigation. Even if followed up, one would have difficulty in determining how important a role one individual or investigation played in influencing an important change.

The altruistic goal of influencing positive change is one of my justifications for finding some otherwise very fine specimens of engineering deficient in some aspect of safety.

The blowmolders mentioned earlier appeared state-of-the-art in all respects except for operator safety. The same could be said for the aluminum extrusion machine, the telescoping crane, and some aerial platforms. In some cases, I have gained the impression that the best engineering went into every aspect of design, except for employee safety.

No longer is it the case that people accept mishaps without seeking compensation from those at fault. Thus, all parties to an accident should be much more introspective and interested in correction, if for nothing more than self-preservation.

4.15 CONCLUSION

It is hoped the reader has gained some understanding about forensic industrial investigations. What has been discussed is only a brief overview of a large and diverse field of endeavor. Although there is much diversity, there is also some similarity in the end objectives of the various investigations.

The objective of all forensic activity should be to uncover the truth. Full attainment of this high-sounding goal is not practical, even in the factual description of the event. We are all witness to the difficulty of absolute determination of infractions at events such as the Super Bowl, even though a host of cameras are trained on the event. Thus, what can be determined at best is the most plausible, most likely, or, preferably, the most probable sequence of events resulting in an accident, based on the investigator's understanding of the facts.

When investigators judge the defective or nondefective state of a machine or product, they are placing themselves on even more debatable ground. They are giving opinions based on their interpretation of the facts and their understanding of sometimes confusing value-judgment criteria laid down by the courts. The investigator can expect that other experts will find some areas of disagreement with their opinions. Ultimately, the validity of an opinion, and how well it holds up under fire, is proportional to the weight of the bases on which the opinion is developed.

As forensic experts, we can only offer and strive to do our best. We hope as a group that something positive eventually results.

Information Sources

The organizations listed below are excellent sources for many books and documents applicable to the field of industrial accident investigations:

National Safety Council (NSC), 425 North Michigan Avenue, Chicago, IL

National Fire Protection Association (NFPA), Quincy, MA

Occupational Safety and Health Administration (OSHA), U.S. Department of Labor, local OSHA area office

American National Standards Institute (ANSI), 1430 Broadway, New York, NY 10018

National Technical Information Service (NTIS), 5285 Port Royal Road, Springfield, VA 22161

American Society for Metals (ASM), 9509 Kinsman Road, Metals Park, OH 44073

American Society of Safety Engineers (ASSE), 1800 East Oakton Street, Des Plaines, IL 60018-2187

Recommended Books

Accident Prevention Manual for Industrial Operations, 8th ed., 1980. National Safety Council, 425 North Michigan, Ave., Chicago, IL.

Gloss, D. S., and M. G. Wardle. 1984. *Introduction to Safety Engineering*. New York: John Wiley and Sons.

Kolb, John, and S. S. Ross. 1980. *Product Safety and Liability, A Desk Reference*. New York: McGraw–Hill.

Metals Handbook, 9th ed., Vol. 11, 1987. American Society for Metals, 9509 Kinsman Road, Metals Park, OH 44073.

OSHA Safety and Health Standards for General Industry (29 CFR 1910), U.S. Government Printing Office, Washington, DC 20402.

5. Product Liability Engineering

LINDLEY MANNING, P.E.

5.1 INTRODUCTION

Product liability—Everyone has heard about it, the newspapers report on it, and legislatures debate it. It is roundly cursed or praised by the most diverse and unlikely people, but few understand it. The subject can be complex, and it can require a law degree or extensive experience to grasp the full range of possibilities and limitations in its application. This short chapter cannot provide that depth of enlightenment, but a brief overview may be useful for several reasons.

This chapter may help the reader to become a better-informed citizen. Most of us are not qualified to act as scholars of constitutional law. We do, however, know the basics, as well as where to find more information (or at least who to call for help). When asked, we are all eager to offer an opinion about legal issues. Many calls for change in the legal system occur in all forms of communication media today, along with demands on legislatures for change. An informed voter should be at least conversant with product liability basics.

Because product liability probably more directly concerns the daily actions of the readers of this book than does constitutional law, some knowledge will be helpful to all readers. This is true for design engineers, for anyone acting to put products into the marketplace, and for consumers and potential plaintiffs, as well as for those who serve as expert witnesses in litigation.

This chapter provides enough information to allow the reader to "read between the lines" of a newspaper article to better understand the real story. It provides an understanding of the terminology required for advanced reading on the subject, along with references for such additional reading. This chapter also discusses several case histories to help readers develop an insight into application of principles to specific cases. In addition, there is some discussion of the myths surrounding the subject of product liability litigation.

Let us begin with a common myth. There never was anyone, from Florida or anywhere else, who was awarded a million dollars for an injury sustained while trimming a hedge with a rotary lawn mower. There was never even a one-dollar award, because there was no such lawsuit. Probably there was never such an injury, but we cannot be certain. If such an injury did occur, the person was probably too embarrassed to report it. The origin of this myth has been traced to deliberate, deceptive advertising.

There have been many incredible stories about the abuse of product liability laws, most of them no more factual than the lawn mower incident. Nevertheless, there have been actual abuses. Most blatant abuses reported in the press are eventually reversed on appeal to a higher court, but not without great cost to the defendant. The unfortunate defendant is often forced to spend a considerable amount of money on defense, and the unrecoverable time of key personnel. Sometimes the defendant asks for trouble, by exhibiting a poor attitude; this is discussed later.

Published reports of a product liability action often make the decision appear either inadequate or unjustified. This applies to outcomes favoring the plaintiff as well as those that favor the defense. In most of these published accounts, either the entire story has not been told, or the story is not really over yet. There is an old saying: "If it looks too good to be true, it probably is." When evaluating the typical media report on a product liability case, it can be said that "If it looks too *bad* to be true, it probably is."

5.1.1 Definition of Product Liability

Most definitions of product liability require a long paragraph or even a page or two. With some fear of sins of omission, a brief definition is offered here:

Product liability is the legal process that puts someone's money at risk because someone else erred in the design or fabrication of an item sold to the public, so that another person was physically or financially injured. This error may have been real, or may have existed only in a person's imagination. The greatest *number* of product liability cases involve injury to a person's body, or what lawyers call "personal injury." The greatest amount of *money* is associated with a small number of "big-ticket" cases involving financial loss to businesses.

The first lesson to gain from this definition is that there are three ways to win, or lose, a product liability action:

1. Deficient design
2. Deficient materials and/or workmanship
3. None of the above; but a lawyer *thinks* one of the above is true

In the last case, it is necessary to help the lawyer see the weakness in the claim. This can usually be accomplished, although not without difficulty.

The first and second cases warrant attacks directed against mistakes in design and manufacture and against the individuals reponsible for the mistakes. Unfortunately, many people expend a great deal of effort to cover up their mistakes. This attitude is common, and is often the best way to ultimately lose the very money the effort is designed to protect. A case history will serve to illustrate this point. A paper prepared on the specifics of this case was published in *Mechanical Engineering*, May 1987.

A few years ago, I was involved in the investigation of an accident in which an elevator fell "up," causing serious and permanent injury to two passengers. Prior to the trial, the injured parties would have settled for less than half of the fabled "million dollars."

Well before the trial, an official of the firm that manufactured the elevator had signed a list of answers to questions that lawyers call *interrogatories*. These questions are official court documents, and false answers may be considered perjury. One of the questions concerned prior failures of the type encountered in this accident. The official assured the court that there had been no other similar failures. Just before the trial, the attorney for the plaintiffs discovered that a year earlier the same corporate official had signed a million-dollar check for an almost identical accident.

The outcome of this revelation in the courtroom was an immediate settlement for approximately three times the amount requested prior to the trial. This manufacturer was indeed lucky to get off so lightly. The jury that heard the revelation would likely have awarded punitive damages of an astronomical nature. The settlement also helped keep everyone out of jail. The local lawyer hired to represent the company knew nothing of the prior accident. He all but got on his knees to proclaim his innocence to the judge. Fortunately, the man's reputation was spotless, and everyone believed him.

It is nice to hear a "good triumphs over evil" story now and then, but there really are not many of these stories, primarily because there just is not that much evil out there. Most manufacturers are as wary of litigation as the rest of us, so they are generally open and honest about most matters (particularly those matters that can be checked). Most do well with the "nothing-but-the-truth" part of their testimony. Sometimes, however, it is difficult to get "the whole truth."

5.1.2   Legal Systems

An understanding of product liability requires some knowledge of the applicable law and how it came to be. The following overview may help. The United States operates on a system of law derived from the constitution and from an evolution of English common law. Most European countries, and many countries elsewhere, use what is known as "civil law." This law is a supposedly complete and detailed book of "statutory law," that is, law written by a legislative body and not subject to court interpretation.

In the United States we do have statutes, but they are open to court interpretation, and even to invalidation. The term *civil law* is used in the United States, but with a very different meaning. In the United States the term refers to legal actions of any form other than criminal law.

The simplest way to understand the difference between civil law and criminal law is by example. The traffic ticket received for speeding involves criminal law. The elevator accident described in the foregoing case study is a civil law case. If the corporate official had been charged with perjury, that would have been criminal law. There are several differences between civil and criminal law, including the fact that an individual can go to jail for a criminal offense, but not for a civil offense. The differences are not particularly important to a discussion of product liability.

Most product liability law is civil law. Criminal prosecution can take place in cases relating to a defective product. In these cases, the prosecution will be centered on the criminal actions of a person rather than the existence of a defect in a product. Somewhere, a prosecutor may have prosecuted a machine (some will try almost anything), but such a case is rare, occurring mainly in movies. With the development of artificial intelligence, such cases may be forthcoming.

5.2   PRODUCT LIABILITY LAW

It has been reported that product liability dates back at least to the Code of Hammurabi, as translated from stone or clay tablets. The translation is there, but it is probably more an interesting comment on the similarity of human conduct in differing places and times than on the basis of our laws. Actually, these tablets were discovered around the turn of the century, well after English common law was established. This common law has served as the foundation for much of the legal system in the United States.

5.2.1   English Common Law

Some of the English law was "statutory law;" much more was what we now call "case law." This law consisted of precedents established by court decisions that had considered similar problems at an earlier date. The courts interpreted the statutory laws, and considered common understanding of the citizens in reaching their decisions. These prior decisions carried great weight in subsequent cases, and added stability to the legal system, as they were infrequently changed.

The key to the success of this system is that it was responsive to the desires of the citizens and to social and technological changes. Changes in the law were often ponderously slow, but they did take place. Meanwhile, the stability of the body of case law prevented rash changes from occurring with each change in government.

Statutes and case law were developed in the English common law to protect the consumer from defective products, even before Columbus sailed for the Americas. Those were simpler days with smaller populations and little travel. Consumers generally purchased products directly from local manufacturers. Because of this, it was expected that there would always be some form of contractual relationship between the manufacturer and the consumer.

5.2.2 United States Law

The English requirement for a contractual relationship, often called *privity*, carried over into the law of the United States. The contract was not necessarily a written contract. As society gradually became more complex, middlemen appeared. Privity was then taken to be the contracts established all along the chain of commerce. If a farmer's plow broke, he looked to his local blacksmith for recovery. The blacksmith went to his supplier for indemnity, and the supplier, in turn, went to the manufacturer for repayment of his loss. Privity kept the farmer from going directly to the manufacturer for recovery, because there was no contract between the farmer and the manufacturer; his transaction was with the blacksmith.

As society and technology grew increasingly complex, the doctrine of privity became unwieldy and less equitable. The courts gradually began to erode this concept. About the time of World War I, a case involving an automobile essentially eliminated the doctrine of privity, if the manufacturer was proven to be negligent. Negligence is an improper act, or lack of the degree of care normally expected of a manufacturer of the product in question.

Another aspect of privity involved warranty, a promise that a product will or will not do something. Another automobile case allowed the warranty expressed in advertising to carry through to the consumer, despite lack of privity. This was not the last use of advertising material against a manufacturer; a later case is reviewed in Section 5.3.2.

This discussion is gradually beginning to use some of the vocabulary of the lawyers. This is one of the most difficult things an engineer must learn in preparing to work within the legal system. At first, it appears that lawyers have rewritten the definitions in the dictionary to suit themselves, stirring in just enough Latin to keep us from having the foggiest idea what they mean. After awhile, one discovers that the vocabulary of the legal system has removed some of the ambiguity of the language, but this comes at the expense of clarity and brevity.

The term we must now introduce is *tort*. This is a bit of the Latin and simply means "a wrong done to someone, such as bodily injury or financial loss." Product liability actions are initiated to recover what was lost because of a tort.

5.2.3   Strict Liability

When most laypersons think of product liability, they connect the term with "strict liability in tort." Strict liability in tort was a small step in the evolution of the law, as it adapted to societal changes.

The law changes very slowly. Courts will go to great lengths to decide a case on the smallest point of law, rather than to rule on some "landmark" aspect of a case. This helps to preserve the stability of the system.

Strict liability was a change to remove the necessity of proving negligence to win a case. Again, an example may help. Suppose a person operated a bottling plant and did a good job of it. The building and equipment were new, maintenance was oustanding, and quality personnel were hired and well trained. Suppose this plant was a showplace for the industry. In spite of all this, suppose that you found a mouse in a bottle of soft drink originating from this plant and you actually became physically ill as a result.

Prior to acceptance of the doctrine of strict liability, it would have been necessary to establish that the bottler had been negligent in order to recover damages. This would require proving that other manufacturers would have operated better than the bottler in question—a very difficult task. But the mouse was in the bottle, and you did get sick! Strict liability implies that if the bottle was defective when the manufacturer shipped it, and if that defect is the source of your problem, then the manufacturer should compensate you for the damages.

If the manufacturer had not shipped the defective bottle, you would not have become ill. Strict liability reasons that there is no more appropriate source to compensate you for your loss than the manufacturer. With today's complex products, the producer of the defect, no matter how much care was exercised, is responsible for an injury caused by a defect.

The effect of strict liability may be considered a form of socialism, in that it spreads the risk of injury from a product around to all users, through costs passed on to the consumer. The effect is much the same as the various forms of insurance purchased to share risk.

5.2.4   Restatement of Torts

The development of law by court decision is cumbersome. The fifty states all have many courts, in addition to the federal courts. Even if applicable decisions were limited to those from state supreme courts, the burden of research necessary to find precedent for each new trial would be overwhelming.

To cope with this task, the American Law Institute created committees to summarize all this material in volumes called *Restatements of the Law*. These provide quick reference. The development of these summaries began in the 1930s, and is an ongoing process. The current portion applying to product liability is known as *Restatement of Torts Second*. This is neither statute nor case law, in the strict sense. However, the prestige of the association and the legal scholars doing the work has made it widely accepted as the best presentation of the current state of the law. And the reference certainly saves everyone a great deal of work.

Product liability, as accepted in the *Restatement of Torts Second*, gained its current interpretation primarily as the result of a landmark case in 1963, *Greenman v. Yuba Power Products*. This case involved a man who suffered an injury when a piece of wood flew out of a lathe. A design defect in the method of tailstock retention was alleged, because the set screws worked loose. The details of the case are widely available, so they need not be given here.

In this case, the Supreme Court of the State of California eliminated all defenses having to do with notice of breach of warranty. The court declared that liability on the part of the manufacturer could be established if the plaintiff could show that the product was being used in the way that it was intended to be used, and that the product contained a defect in design and manufacture of which the plaintiff was not aware, which made it unsafe for its intended use.

5.2.5 Modifications to Strict Liability

Modifications to the foregoing decision have been made as the need has arisen in various courts. The phrase "use for which it was intended" has been extended to "foreseeable misuse." One example was the case *Larsen v. General Motors*, which involved the concept of "crashworthiness." The defense contended that involvement of a vehicle in a crash was not what the vehicle was sold for, so that the manufacturer should not be held liable for any injuries caused by hazards in the way the car treated the driver during a crash. The court determined that a manufacturer should design a product to meet any emergency use to which it may reasonably and foreseeably be put.

Crashes for automobiles are surely foreseeable by anyone. Since that decision, substantial improvements have been made to automobiles to protect occupants in collisions. In fairness to the manufacturers, some improvements in crashworthiness had been made earlier, and others would have been made even without the decision. It just is not good business to maim one's customers, nor are all automobile corporate executives ogres; most have families that they put into their own products, so they also have a personal interest in safety.

Foreseeable misuse does not extend to hedge clipping with a rotary lawn mower, but the courts are constantly deciding how far it does extend. I recently worked on an injury case in which a product had actually malfunctioned. In this case, the jury found for the manufacturer. The case involved a snow blower that was attached to a garden tractor. Some set screws had come loose, permitting a belt to jump off a pulley into a position where the clutch could not release the drive on the snow blower. The jury decided that there were enough warnings to shut off the engine, and that the operator was sufficiently skilled, so that he should not have relied on the clutch to work the way it was intended. The case may be appealed.

5.2.6 Failure to Warn

Warnings are frequently provided with products, and improper warnings and failure to provide warnings are often cited as defects in a product. In one of the preceding case studies, jurors thought the warnings were adequate and that the injured person should have obeyed them. Other juries may not have reached the same conclusion.

To be effective, warnings must be visible in obvious locations, and must clearly warn of the specific hazard involved. Warnings may not be used as disclaimers, and will not be so accepted by the courts.

5.2.7 Subsequent Change

English common law was carried over to U.S. law in one other area worthy of note. Society has considered it preferable to repair a hazard, rather than to leave it in place to injure others. In keeping with this philosophy, it has not been possible to use the argument in court that something must have been hazardous because the person in charge changed it to a safer condition.

In strict liability cases, this concept has recently been modified. The California Supreme Court, in *Ault v. International Harvester*, found that such evidence is admissible, but only if the plaintiff can show that the changes were made to correct a prior defect, not merely to update or improve the product.

5.2.8 Defect without Liability

Another interesting recent case involved a tube-and-wire supermarket shopping cart. The single jury verdict on this case was rendered less than a month prior to this writing.

A woman was pushing the cart loaded with groceries when, without warning, the cart collapsed, causing minor but long-term injuries. Examination of the cart revealed fatigue failures in the original welds, as well as fatigue failures in repair welds. The repairs had been contracted out by the supermarket to a specialist in cart repair. The repair welds were not well designed, nor were they of high-quality workmanship.

Investigation into this case revealed that there are many brands and types of shopping carts on the market, and they contain substantial differences not readily noted by the casual observer. In testifying at the request of the supermarket, I gave the opinion that the design and manufacture of the cart were defective, and that the repairs to the cart were also defective. In addition, it was my opinion that the fatigue cracks would have been small and difficult to see before the actual collapse of the cart. An engineer or a mechanic would probably have been the only person likely to have noticed the cracks. Supermarket employees would not likely have had the experience to know that the cracks could lead to collapse, even if they had noticed them.

The plaintiffs had brought the suit against only the supermarket, claiming a defective cart, and asking for a jury instruction for strict liability. Representatives for the supermarket responded that they had not manufactured or repaired the cart, nor had they sold it. The defense claimed that the supermarket was not strictly liable and that this was a case of premises liability. To find the defendant liable, the defense claimed, proof of negligence would be required. The defense further claimed that the testimony regarding the difficulty of seeing the cracks and the good-faith attempt to keep the carts in good repair demonstrated that the supermarket was not negligent.

The judge agreed with the supermarket that the defense was not strictly liable. The jury also agreed with the defense position that the supermarket was not negligent. This case may not be appealed, as the amount of the damages was not large; however, it would be interesting to see the outcome of an appeal. The theory that prevailed in this case may not be unique, but I have not encountered a similar case in my practice.

5.3 PRODUCT LIABILITY OTHER THAN STRICT LIABILITY

5.3.1 Negligence

Various court decisions have eliminated the need to show negligence to recover damages for a tort. These decisions have done little to eliminate the use of negligence as an argument, if it can be proved. If a plaintiff can show negligence caused injury, the plaintiff is still as entitled to recovery as in the past.

It is usually easier to show the existence of a defect, and thus rely on strict liability, than it is to show negligence. Because negligence usually takes place in the seclusion of a factory, it is quite often very difficult to prove. An informer helps. Unhappy ex-employees have been known to provide the needed information. Also, "whistle blowers" motivated by a sense of justice may come forward. More luck than that involved in picking Derby winners is required to find the right person at the right time, but sometimes that person can be found.

The manufacturer is much more cooperative in helping the investigator find defects, than in helping establish negligence. The manufacturer obligingly sells the investigator any number of the offending products, to be dissected at leisure and to reproduce the conditions leading to the injury. All the examples the investigator can afford to buy are usually available for experimentation.

5.3.2   Breach of Warranty

Breach of warranty is a legal avenue available to an injured party in some cases. If the warranty is written, either in the form of a specific document provided with the product or contained in advertising material, the proof of breach may be easier than in the case of an implied warranty. In the latter case, it must be shown that the act of selling the product implied certain things about the fitness of the product for use and the way it was expected to perform the desired task. Warranty may be implied simply by the act of selling the product. A reasonable person should expect that the product will safely accomplish the purpose for which it was sold.

Advertising formed the basis for a major change in the law, as previously discussed. Advertising has also been used as evidence to support the claim of a product defect. The numerous cases involving Jeep vehicles used the company's own television advertising to show that safety was less important to the manufacturer than selling the product. Some of the TV advertisements depicted Jeeps, usually operated by young people of either sex and without special safety equipment, flying through the air with the greatest of ease, as they cleared the tops of amazingly steep hills.

The attorneys for the plaintiffs had a field day with this advertising, claiming that this deceptive advertising led the unwary customer, particularly the young customer, into the false belief that the vehicle was capable of performing such feats with impunity. These advertisements also implied that the vehicle was safe for the less strenuous act of driving on the highway, at whatever speed the vehicle could sustain.

Testing revealed that the vehicle did not perform well at highway speeds. The results of the testing confirmed the predictions many engineers made by simple observation. Jeeps, particularly the CJ-5 model, were found to be directionally unstable when encountering a minor disturbance in their path. They slid sideways easily and became very demanding to control. The CJ-7 model, with a longer body, was somewhat safer, presumably because the extra length increased the moment of inertia in yaw, spread the point of application of control forces farther from the center of gravity, and increased the time available for the driver to correct directional deviations.

Testing also revealed that when the Jeep turned sideways to its path of motion, it would overturn, even at very low speeds. The violence of the overturn in some cases was surprising. In the 30-mph sled test conducted for Jeep by General Motors, a Jeep was placed sideways on a wheeled, inclined plane sled at 15° to the horizontal. The low edge of the plane faced the direction of motion, as did the low side of the Jeep. The sled was suddenly stopped from a speed of 30 mph, allowing the Jeep to roll forward. As expected, the initial roll was onto the side and up onto the roll bar. From this point, the nose went down and the tail up, as the roll continued end-over-end, along with a spin about the longitudinal axis. At one point, the front bumper was down, but about 3 feet above the pavement, with the rear directly above the front. Obviously, in this position, great forces would be exerted on any passengers. Many of the actual Jeep accidents were found to have followed this pattern of vehicle movement.

Some of the earliest Jeep suits related to the quality of design and construction of the roll bar. Initially, roll bars were "after market" accessories sold and installed by Jeep dealers and others, but with the knowledge of the Jeep manufacturer. Later, Jeep produced and installed bars of similar design. Testing revealed the amazingly poor strength of these structures. One early design was found to be so inadequate that the mount used to attach the bar to the body sheet metal was one fourth the strength of the bar.

Gradually, as the model years progressed, the mounting strength and other aspects of the roll bar were improved, but injuries continued to occur. More testing revealed that forces on the occupants during a rollover would permit ground contact for the occupants' bodies, including their heads, even if the roll bar were not crushed, and no matter where the roll bar was placed. This condition was caused in part by the weakness of the windshield frame and the hinges and latch which permitted the windshield to be folded flat on the hood. Full-roll cages with solid or net tops, such as used by professional racers, in conjunction with upper body restraints were needed to provide any significant improvement in occupant protection. The best protection was still to keep the Jeep on its wheels.

The defects in the Jeep were cited so often they were categorized, with proper selections made to fit the case at hand. Thus, a Jeep case was "a handling case," "a propensity to overturn case," "an occupant protection case," or "a failure to warn case." Some cases involved all of these. These defects were well explored in many cases; probably Jeep alone knows how many. As a part-time "expert" with a full-time teaching position, I testified in a dozen cases, myself. At one time, one of my colleagues was involved in over 80 active cases and had been involved in many others that had been settled. The volume of cases indicates that this product should probably never have been sold to the general public.

Despite all this, the Jeep was probably one of the best special-purpose vehicles available. It was not a safe highway vehicle, but it was very mobile on rough terrain. I have used a Jeep to snake logs out of the forest, and then to tow a trailer loaded with a ton of firewood. The Jeep presented no problems in the woods, and actually felt more stable on the highway towing the loaded, two-axle trailer.

Jeep advertised the way they did in the apparent belief that such advertising was required to sell vehicles. The corporation defended the lawsuits as if the corporation believed the vehicles were safe for any ordinary use. Once this path of defense was chosen, it would have been difficult for the corporation to take another position. However, a different course of action may have sold more vehicles, and perhaps may have saved Jeep most of the money lost in the suits.

Jeeps had a "macho" image. This image could have been used in the advertising to indicate that design compromises had been made to adapt the vehicle to the roughest of off-road applications. The ads could have pointed out the dangers of using the vehicle on the highway. The implication would have been that "only the best could control the beast." Buyers would have been made aware of the danger, and would then have assumed most of the risk. There was no way to remove responsibility for poor engineering of the collapsing roll bars, so these suits should have simply been settled. There are clear lessons for other manufacturers in this case study.

5.4   OTHER LEGAL DOCTRINES

5.4.1   Comparative Negligence

The doctrine of comparative negligence is applied in other actions between persons. For example, in an automobile accident where one party is judged to be 60 percent at fault, and the other party 40 percent at fault, the party who is 40 percent at fault will have damages reduced by that same amount. This rule has been applied in some jurisdictions in product liability cases. In other jurisdictions, this doctrine is not applied. To further confuse the issue, some jurisdictions, such as the State of Nevada, require that the plaintiff prove the defendant is at least 51 percent at fault before the plaintiff can recover any damages. Other states, such as California, do not have this rule.

5.4.2   Deep Pockets

The doctrine commonly referred to as "deep pockets" is formally known as "joint and several liability." Suppose that several defendants have been found at fault for an injury to a plaintiff, and each has been assigned a fraction of responsibility. The dollar value of the damages has also been assessed. If each defendant can pay the assigned share of the award, everyone goes away with the judgment satisfied. If one or more cannot pay, then the doctrine of joint and several liability requires those remaining who have money to pick up the whole tab.

Some states have eliminated this doctrine on the theory that it is not fair to ask a defendant who has only a small responsibility for an injury to pay the entire damages; other states continue to employ this doctrine. Presumably, the reason for joint and several liability is that, absent the fault of each of the defendants, there would not have been an injury. Thus, any one of the defendants who are able should make full restitution, if necessary.

5.4.3   Workers' Compensation

Most, if not all, states and other jurisdictions have enacted workers' compensation insurance laws. These laws provide for medical treatment and some monetary compensation for a disabling injury. In return for contributing premium payments to these insurance systems, employers are protected from suits brought by injured employees. Some employers, such as some railroad corporations, are not included in these insurance programs, and thus may be sued by employees. These suits would likely be negligence suits, rather than product liability suits.

In most cases, compensation by these programs is considered by most persons to be inadequate. One potential recourse to an injured employee arises if the employee was injured by a defective product not produced by the employer. The manufacturer of such a product is not protected by the workers' compensation laws, and may be sued. Many such suits are filed, and many are decided in favor of the injured employee.

5.5   COURT SYSTEMS

Thus far in this discussion, several types of courts have been mentioned. An understanding of the relationship among these various courts and some idea of their operation should be useful. I am not an attorney, and this discussion is based on my own personal experience. My experience has been that there are substantial differences between court systems. Thoughtful strategy is required to select the right court and to prepare an appropriate case presentation. A good attorney is well versed in the differences between courts.

5.5.1   Federal Courts

Most larger cities have a federal district court, which is the entry-level court in the federal system. Judges serve for life, unless they are convicted of a serious crime. These judges hold an exalted position and an extraordinary degree of power. The attitude of district court judges toward their own position of authority is legendary. It is extremely important to show respect toward these judges in the courtroom.

This first level of federal court is the place where the actual trial takes place. If a jury is used, it consists of twelve persons. To reach a verdict, they all must agree. State courts often use only six to eight jurors, and require only a two-thirds or three-quarter majority vote to reach a verdict in a civil case.

The federal rules of evidence, discovery, and procedure are precise and are strictly enforced. State courts have rules based on federal rules, but they are often more flexible and less stringently applied. This often prompts plaintiffs' attorneys to go to great lengths to keep a case in a state court. Defense attorneys often try to get cases moved to federal courts. The appropriate selection of court will be discussed further.

There are two levels of appeal courts above the district courts. The first court of appeal is the circuit courts, situated in several locations around the country. Circuit courts consists of panels composed of judges who hear appeals based on claimed errors in procedure or similar technical arguments. These courts do not conduct trials or hear new evidence; they are essentially courts that review what has gone before, hear arguments, and vote on the dispute. The circuit court may confirm the decision of the trial court, or reverse the decision and/or send the case back for a new trial. The circuit court also has additional powers that generally do not apply to product liability cases.

The U.S. Supreme Court is a well-known component in the legal system. This court is much like the circuit courts, but is considered the final authority. It is rare for a product liability case to reach this august body, but it is also rare for any kind of case to get this far. Those cases that do reach this level of adjudication are those that establish important precedent.

Tax courts and other administrative law courts also function within the federal government. Some of these courts handle technical matters and manufactured product disputes, and use technical experts to arrive at establishing fines to be levied against manufacturers. The work of these courts, however, is not considered product liability. These courts deal with the enforcement of federal regulations and not the awarding of damages for torts.

5.5.2 State Courts

There are many types of state courts, and they differ in form more than in substance. Some insist on exactly the right color of ribbon used to tie each scroll-like document before it is presented to the court, just as the old English courts did. Others are so modern they resemble an informal mock trial in a classroom. These differences exist not only between states, but also between counties in the same state.

Some states have a three-tiered system, much like the federal courts. Others, such as Nevada, have only two levels. The systems do monitor developments in other jurisdictions. Decisions in cases from neighboring states may be cited to support a position. There are important differences, such as the time permitted between an injury and the filing of a suit by a plaintiff. Nevada permits 2 years to elapse, whereas its next-door neighbor, California, permits only one year. Exceptions are provided for special cases. For specific details, an attorney in the appropriate jurisdiction should be consulted.

In some states, the process of arbitration is used to settle civil suits. These are usually negligence cases, such as to establish the party at fault in an automobile collision, and are not likely to involve product liability. Attorneys and expert witnesses may be involved, but the monetary damages sought are usually too small to warrant the preparation required to discredit a product. Small claims courts are also unlikely to hear product liability actions, for the same reason.

Other courts, such as the justice of the peace and municipal and traffic courts, are generally criminal courts, where product liability is not mentioned. The exception might be the case where the existence of a product defect could be cited to exonerate a defendant.

An anachronistic aspect of the legal system used in some states is "sovereign immunity." In the old English law, it was impossible to sue the king. There are no kings in the United States, but at least one state, Nevada, employs a similar concept. Although allowing itself to be sued, Nevada limits the amount of damages to $50,000, no matter how serious the grievance. It is unlikely that a state, even Nevada, will be in the business of making products for sale, but the state does construct various public works, such as fences which trap wild mustangs on a highway.

Earlier, it was noted that lawyers try to get their cases into the desirable courts. This is not always easy. There are rules regarding which type of case belongs in each court. However, there may be some flexibility to exploit the system. If an injury occurs in a state where the defendant has a place of business, the trial takes place in the courts of that state. If the defendant does not have a place of business, but the product was sold by a dealer in the state of injury, and if the dealer is also being sued, then the case will probably remain in the state court. If the injury takes place in a state where the defendant does no business at all, the trial will probably be held in federal court. Naturally, if a federal agency is a party to the suit, the case will be tried in federal court.

A suit may be filed in the state of incorporation of the defendant, even if the accident takes place in another state. For example, American Motors–Jeep was a Nevada corporation. Thus, the courts of this home state for the corporation became the site of the numerous Jeep trials discussed previously.

The reader should bear in mind that all of the foregoing comments are from an engineer's perspective. An attorney should be consulted regarding legal subtleties that may influence the appropriate court for a specific case. In addition, the law is constantly changing, and a good attorney expends much effort to keep informed of revisions as they occur.

5.6   THE LITIGATION PROCESS

If there were no injuries caused by products, there would be no lawsuits. The purpose of a lawsuit is to recover damages, either physical or financial. If all the industries of the world were to produce perfectly designed and manufactured products, this chapter would become obsolete. Impossible? Of course. But perfection is a reasonable goal, and it can be approached asymptotically.

5.6.1   The Manufacturer

Experience, rather than any statistical study, has shown me that the manufacturers who conscientiously seek to design and produce safe products come very close to succeeding in the attainment of perfection. Some manufacturers are rarely sued, and when they are, the related injury is normally less serious than those suffered by users of a competitor's product. One of the reasons is that some manufacturers simply "make a better mousetrap"—better from the standpoint of safety. Not surprisingly, these safer products also seem to last longer and to require less maintenance. Sometimes such products cost more, but usually the wise consumer chooses the better product, even if it is more expensive.

Another reason some manufacturers appear to be "judgment proof" is their attitude in conducting good-faith negotiations with the injured party. These negotiations include admitting fault if it appears that fault is present. The attitude that "there can't be anything wrong with our product" may prove to cost the manufacturer a great deal. The case history of the elevator accident in Section 5.1.1 illustrates this point. This attitude cost the manufacturer three times the amount it would have cost to admit fault prior to the trial. Worse than that, discovery of the denial of the previous accident was the reason for the large settlement. Simply admitting that a similar accident had occurred before would have saved the manufacturer a lot of money. At the time of the second accident, the manufacturer was working on ways to solve the problem that had caused the first accident. Had the manufacturer discussed this openly, the earlier failure could not have been used against him to the degree it was when he lied about it. The attitude of the manufacturer toward potential product defects is an important factor in determining the outcome of a suit.

Suppose the opposite situation develops. A conscientious manufacturer makes a good product, but someone is injured while using it. The manufacturer makes a thorough study of the situation and the facts, as the manufacturer understands them, indicate that the product performed flawlessly and that the user was at fault. What is the likely outcome of the suit?

Almost certaintly, the manufacturer can win the suit, if a good attorney is retained. A poor lawyer can lose the best case. The stature of the lawyer for the plaintiff usually is not quite so critical, assuming a minimum level of competence. There is a limit to what the best lawyer can do, if properly opposed by a competent lawyer. The law is really more specific than most persons believe, and when a pair of good lawyers oppose each other in a case, they usually conclude the case just about where the law says they should. There are occasional exceptions, but they can go either way.

Some manufacturers, such as the elevator manufacturer in the case study, refuse to settle a suit before trial; others will consider a "nuisance settlement," a settlement payment less than the cost of defending the suit. Still others, such as the Jeep manufacturers, have lost so many costly suits that a data base has been developed to reasonably evaluate suits for settlement.

5.6.2 Plaintiffs

Plaintiffs are like number systems: real and imaginary. A career military officer, with a master's degree, who becomes a quadriplegic with brain damage as the result of an accident is about as real as they come. The man who had his friend cut off part of his leg with a chain saw, to fake an accident to collect insurance compensation, represents the other end of the spectrum. Both examples are actual cases.

Most injured people do not know how to go about protecting their rights and recovering their damages. They often inadvertently destroy evidence, sign releases prematurely, and make incorrect admissions against their own interest. Sometimes, unscrupulous claims adjusters apply pressure in these directions, although most claims adjusters are ethical in their dealings with injured parties.

The first step for an injured party to take is to retain a lawyer. The lawyer should be carefully selected based on the specifics of the case. The lawyer who handled Uncle Joe's estate or Aunt Sue's divorce is probably not the right one, just as a plumber would not be the right person to fix a wiring problem. A specialist is needed. Maybe any personal injury lawyer will do, or maybe one is needed who knows all there is to know about chrome-plated, left-handed monkey wrenches, or Jeeps, for that matter.

Often, all aspects of the plaintiff's life have been demolished by the accident, including emotional, family, physical, and financial resources. Sometimes the attorney is the only one around who can provide for the care of these aspects of the client's life. The military officer mentioned earlier had a wife who did all she could to care for her husband. Despite having full government financial support, she was holding onto her emotional health with nothing more than grim determination. Many such persons simply cannot endure the added strain and indignity of a trial. Sometimes just the interminable delays encountered before a trial are beyond their capacity. The good attorney knows the client well enough to understand these situations. Sometimes a settlement is best, even if it does not result in as much compensation as the case would otherwise warrant.

Most attorneys take a case on a contingency fee basis. This means they do not get paid for their time unless they win the case. If they win, they usually receive about one third of the settlement, after expenses. If appeals are involved, they may receive half of the settlement, or more. The benefit of this system is that at least one attorney must think the case is good enough to bet his or her salary on the outcome.

5.6.3 Experts

The attorney must select the experts needed to present technical evidence to the jury. If the right attorney is chosen, the right experts will usually be instantly available. If not, some difficulties may arise. The wrong expert may destroy the case. One of the worst things that can happen is to retain an expert who tells the client just what the client wants to hear. If this expert is proven wrong in front of the jury, even a good case is not likely to survive.

If the case is large enough, or complex, more than one expert may be used. Usually the work of these multiple experts overlaps, but each brings some unique training or experience to the problem. These views from differing directions are likely to reveal flaws in the theory of the case, to allow for a more complete presentation and to avoid embarrassment in front of the jury. The disadvantage of multiple experts is the expense.

Some agencies offer lists of available experts, for a fee. These agencies operate much like employment agencies. Usually no screening is done regarding the expert's qualifications. Books of experts are also made available by law book publishers. Most of these experts are not screened. The only requirement is for the expert to pay a listing fee. Some technical societies provide similar lists, but at least the person is screened to the degree necessary for membership in the society.

There are societies consisting entirely of experts. One of these societies is the American Academy of Forensic Scientists. Its members are persons allied primarily with the medical professions; however, an engineering division has recently been added. Another society, the National Academy of Forensic Engineers, consists entirely of registered professional engineers who are well screened prior to admission (see Section 1.4.1). Expert witnesses selected from these two societies would be appropriate for most product liability actions.

5.7 TOOLS OF THE TRADE

Unfortunately, it is not possible to give a shopping list for the tools needed by a product liability engineer; the field is too diverse. I could suggest a scanning electron microscope and a ten-mile test track for starters. I have neither, but I do know where to borrow them. Product liability engineering is clearly not a single discipline; it involves all engineering disciplines. Forensic practice requires a specialist from the proper field for each case. The guideline should be that the expert is engaged to search for the truth in an unfortunate event, and then to present that truth in a manner that can be understood by laypersons. This is a challenging task.

References

Barzelay, M. E., and G. W. Lacy. 1964. *Scientific Automobile Accident Reconstruction.* Albany, NY: Matthew Bender Co.

This reference is a set of six volumes, written for lawyers. The extensive compilation of information includes accident reconstruction and product liability relating to vehicles. The sample testimony included is excellent for training an engineer in how to present evidence in court, how to handle cross-examination, and how to work with lawyers and the legal system. This reference is revised and updated regularly.

Ross, K., and M. J. Foley, 1980. *Product Liability of Manufacturers: Prevention and Defense.* New York: Practicing Law Institute.

This 1000-page handbook was written to accompany a course for lawyers. It addresses the law technically, but is readable and is of benefit to engineers and others interested in the field. This reference is also revised regularly.

Smith, C. O. 1981. *Products Liability: Are You Vulnerable?* Englewood Cliffs, NJ: Prentice–Hall.

This outstanding 300-page volume covers the law of product liability in full detail, with many citations. Preventative measures are discussed and numerous case histories are included.

Thorpe, J. F., and W. H. Middendorf. 1979. *What Every Engineer Should Know about Product Liability.* New York: Marcel Dekker.

This 100-page volume covers the law and reduction of exposure. It examines insurance, litigation, ethics, and the role of engineering education.

6. Traffic Accident Reconstruction

JOEL T. HICKS, P.E.

6.1 INTRODUCTION

The forensic engineer practicing in the field of traffic accident reconstruction attempts to provide insight into the essential character of such mishaps. The field has been recognized by the legal system for a long time, since at least 1920 (Williams 1982). In that Wisconsin case, which involved a truck that struck a pedestrian who was trying to board a streetcar, the court admitted into evidence an opinion based on skid marks.

Those who wish to know about such things frequently are not skilled in the sciences, and do not easily comprehend the nuances of natural laws. Hence, it is not surprising that successful engineers in this field have a good command of language and an understanding of legal concepts, as well as expertise with dynamics. They are interpreters of one set of laws for those who are familiar with another set of laws.

6.1.1 Typical Clients

Those who seek reconstruction information are varied, as are their reasons for wanting the information. An attorney may want to know if the bizarre story told by his client has merit. A defendant's attorney may want to know if the plaintiff contributed to the seriousness of the affair. Insurance companies typically want to know what caused an accident so they can settle a claim, determine if the policy covers the loss, or determine if their insured is responsible for the damage. Companies whose drivers are involved in accidents frequently just want to preserve physical evidence in case one of the parties decides to file a suit later. The most unfortunate client might be the individual who has become embroiled in a criminal matter as the result of an accident.

6.1.2 Typical Information Sought

The data most often sought relate to the speed of one or more of the vehicles involved. A layperson understands well the common perception of fault resulting from speed in excess of posted speed limits. It is a straightforward concept that requires no special technical expertise. What laypersons are not likely to understand well is the effect of a small *change* in speed on the results of a collision. A difference in speed of 5 mph may mean the difference between a "near miss" and devastating injuries. A question regarding speed might be answered with more than a range of numbers; the proper answer may be an analysis of what the speed means in terms of collision results. Hence, the forensic engineer should expect to devote some effort toward answering questions that the client may not have asked.

Among other common information sought is the position of the vehicles at impact, or their paths or directions prior to impact. The issue is not always as simple as "who was on the wrong side of the road?" Who was on the wrong side of the road *first* may be more important. Similarly, a charge of manslaughter may hinge on whether the vehicle struck was parked 4 feet away from the traffic way or partially in the traffic way. Whatever the client wishes to know is almost always suggested by a cursory review of routine data, such as a police traffic accident report, or by statements regarding the mishap.

6.1.3 Scope and Purpose of Investigation

It is not uncommon for an investigation to change direction in midcourse. For example, a more detailed mechanical inspection of one of the vehicles may be appropriate when an analysis of data suggests that it behaved in some surprising way.

Regardless of the information initially sought, predetermining the scope of a particular investigation is often difficult. It is expected that a client may not know before an investigation which issues may be relevant. It is also true, however, that the experienced forensic engineer may not be able to anticipate all relevant issues either. There are no specific rules to follow in establishing an investigation. Flexibility is required to find the one overriding target, the truth, as facts are uncovered.

The purpose of traffic accident reconstruction is to respond to the quest of the legal system for judgments based on reasonable information. Hence, litigation, or the threat of litigation, is the justification for most of the work. It is amazing, however, how few cases actually reach the courtroom. Perhaps as many as 20% progress to the stage of a deposition by the investigator, and as few as 5% are finally decided by judge or jury. It seems that when conflicting parties are made to understand what the facts imply, they find a way to settle their differences. The forensic engineer can play an important role, outside the courtroom, in assisting resolution of conflict.

Civil law might be the first arena that comes to mind when traffic accident reconstruction is discussed, but this is actually not the primary application. There are far more law enforcement officers with reconstruction training than there are engineers who practice in the field. The thrust of these law enforcement investigators is the support of criminal charges in traffic accident cases. Their training also is a response to the need for a reasonable basis for decision making in the legal system.

6.1.4 Resources

Appropriate references are listed at the end of the chapter. Several of the references have extensive reference lists of their own. The list is not complete. A search is likely to identify several hundred titles.

Law enforcement officers who have training in reconstruction often provide consulting services, and the materials that have been prepared for their training are good resources for the engineer (Baker 1975; Rogers 1981). The engineer must frequently rely on the skill of a police officer to provide information about an accident, and knowing what the officer is likely to have been told helps to put the accident in proper perspective. The text material in these resources usually is a good source of research data, such as coefficients of friction and reaction times, but the mathematics provided often result from simplifying assumptions that are not always explained in the text.

People are good resources, too, and their employment is often beneficial in forensic work. Some practicing engineers leading an investigation are reluctant to admit they do not already know all the answers. This is an unbecoming professional attitude that flies in the face of common sense. No matter how much mechanical experience an engineer has, he or she can usually benefit in knowledge and skill by working beside an experienced mechanic. The same is true of other trades as well. What is important is that the forensic engineer exercise control over the direction of the inspection, and that the technician be taught beforehand the need for proceeding slowly enough to allow the engineer to provide direction. The chain of evidence must be preserved, too, which is another requirement that is typically foreign to the technician. It is the engineer's responsibility to see that this is done properly.

6.2 INVESTIGATIVE TECHNIQUES, PROCEDURES, AND TOOLS

Investigations are just as different and just as similar as people are. Each situation has in common with the next a desire to appropriately relate speed, distance, and time in a way that is meaningful for unique

FLOW CHART

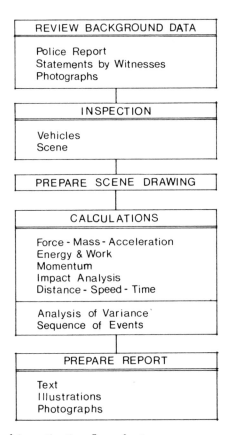

Figure 6.1 Typical investigation flow chart.

circumstances. Although each investigation begins with an inspection, the information sought is likely to be influenced by a knowledge of the analytical tools available and the type of report that is to be made.

A flow chart for any particular investigation is likely to be complex, with many feedback loops. However, there is a common path for most investigations, which will form the basis for discussion of a stereotypical situation. This simplified flow chart is given in Figure 6.1.

6.2.1 Background Information

If background information is readily available, it is usually better to review it before inspecting the scene or the vehicles. The data usually include an investigative police officer's report, and may additionally include statements by the drivers and witnesses. These records may alert the investigator to peculiarities that may have become obscured over time. Photographs taken soon after the incident are of particular benefit.

6.2.2 Photographs

The investigator should make inquiries to determine if photographs were taken. The obvious people to ask are insurance adjusters, the police (particularly if more than one officer was present and if a fatality was involved), and witnesses. Witnesses may remember a bystander having a camera. The drivers and/or their families may think to take photographs; typically they do so the day after the accident.

Photography is an important skill for an investigator to acquire. The camera is a tool that helps others to see what the investigator sees and, therefore, helps others to understand the significance of what is seen. The equipment does not have to be elaborate, but it should be of reasonably good quality. A well-equipped camera bag will include a macro lens (for recording small objects as close as 6 inches from the lens) and an electronic flash unit that can be used with the macro lens.

There are two good reasons for taking photographs: to illustrate a dissertation, and to aid the analyst. A photograph may serve both purposes, but a series of photographs, beginning close up and moving progressively further away, may be required to serve both needs. Plan and elevation views of vehicles help the analyst, but views from each end of a diagonal are usually better for general descriptions in the report.

6.2.3 Inspecting the Scene and the Vehicles

There are times when a scene or the vehicles should be immediately inspected, without a prior review of background information. Such is the case when there exists the possibility that evidence could change, or even be lost altogether. Examples are vehicles that have been sold and that may be altered, or a scene that is likely to be altered by a highway department repair crew.

Barring such a situation, or perhaps adverse weather conditions, it does not matter which is inspected first, scene or vehicles. The data collected all go into a reservoir of facts from which the analyst draws as needed. An investigator frequently returns to a scene after inspecting a vehicle, to confirm that observations at both places are consistent. An example is the record of pavement damage at a scene. If evidence on a vehicle suggests alternative scenarios, the scene should be visited again to determine which scenario is more probable and which can be eliminated from the list of possibilities.

6.2.3.a *Inspecting the Scene*

The objective of a scene inspection is to collect data for analysis and for use in preparing a drawing. Collection of data is important, but it is not the most important thing. As an investigator, you are the most important thing on the scene. You are out of your element, just like a fish out of water. Safety must be of paramount concern, both for yourself and for passing motorists.

One of the rules of conduct is "There should be no surprises." As an investigator, you must be aware of what is going on around you all the time. If you cannot pay attention to traffic and record data simultaneously, employ someone to stand watch for you who understands the gravity of the situation. The same rule should be applied in warning motorists; you must not provide a surprise for them. You do not want to force them to make any decisions regarding your welfare or the safety of your partners. You should make all decisions for them, prior to their arrival on the scene.

Motorists should be aware of your presence before they get close to your position. If the shoulder of the highway is wide enough, park your vehicle so that it does not obscure you, and leave its emergency lights flashing. Weighted triangular markers placed on both sides of the road help to alert traffic. You should wear clothing that contrasts with the background seen by the motorists, and you should not make sudden moves toward a vehicle as it approaches.

The survey methods selected for the scene should afford reasonable security, and your concern for safety implies that you will be interrupted time and time again. It is best to lay out a baseline along the edge of the roadway, and use it to measure each feature, one at a time, as the traffic allows. Choosing a baseline location in common with the police officer's report facilitates cross-referencing.

The best measuring device is a 50-foot, fiberglass surveyor's tape mounted on a hand reel. A wheeled measuring device is also handy. Many police officers use them, and having one yourself enables you to verify measurements. A large-wheeled device is better for rough terrain.

The baseline increments may be marked with a can of spray paint. Be careful not to alter features, such as skid marks, by trying to enhance them for photography. Paint marks may be made nearby for reference, but the evidence must not be altered.

If the scene involves a curved roadway, you may have to find the radius or radii with the use of a transit. Be sure to place the transit on an offset so there is a comfortable distance between you and the traffic. Otherwise a Cartesian, or quasi-Cartesian coordinate system is generally best, where every measurement is taken from a baseline. If you have experience with an alidade or other surveying equipment, these may be helpful; however, simple tools and methods are better with traffic passing close by at highway speeds.

Whether or not a notebook is required for field notes is a matter of preference for the individual. Some find a permanently bound book awkward to use in the field; others do not. A standard tablet may be dedicated to one project, and it may be combined with a clip board and rubber band so that it can be handled satisfactorily in moderate wind. Organization and clarity of the data are more important than the media used. There should be a cover page or two, depending on the length of highway involved, on which to record principal items of interest. The details should be recorded on other pages at larger scale, so that lines and notes are not confusing. These detail pages should be referenced on the cover page(s).

Each feature that may be associated with the accident should be noted and located. If its relevance is not obvious, note it so. A finished drawing should include pavement striping that is as close to the actual scene as possible. Center stripes are almost always laid in 40-foot periods. The length of solid line and space varies, but the period is usually relatively consistent.

If the police accident report refers to a point or area of impact, the investigator should try to locate it and include it in the coordinate system. The same should be done for each feature referenced in the police report. The outside edge of the shoulder is usually a satisfactory limit for the drawing, except for areas involved with the travel of one of the vehicles. Anything that might be associated with a secondary collision should also be located in the data.

A clinometer (level bubble device) will usually provide satisfactory slope and crown values. If the situation requires a surveying level, care should be exercised while handling the rod around moving traffic. The investigator is, again, a "fish out of water."

Everyone wants to know the best, easy way to find an appropriate value for the coefficient of friction, C_f. "Best" and "easy" are not compatible in this case. There are two ways to determine a reasonable value for C_f. The easier is to accept a range of values from published material (see Section 6.3.2.g); the other way is to test the surface.

The best test requires the investigator to "burn some rubber." This may involve test-stopping a similar vehicle under similar conditions in the vicinity of the accident scene. This can be dangerous, and the results may not yield the desired information. If it is decided to test-skid a vehicle, and the speed of the vehicles in the accident was too high to duplicate safely, the test may be run at 30 mph. There exist better empirical data to extrapolate tests at that speed than at others. Also, the investigator should study the techniques and pitfalls involved in such tests; if he or she has never performed a skid test, experienced investigators should be consulted.

An alternative is to construct or purchase a device that has a tire rubber surface and that will measure a pull force that is related to the normal force. One such device is a spare tire that is pulled with a surveyor's spring scale. Corrections must be applied for the slope of the pull line (which cannot be avoided) and for speed. Another test device might be a section of an old tire that has been filled with concrete. If the latter is used, care should be exercised to attach the pull line low on the tire. The moment created in the pull causes the device to tilt and shift the center of gravity. In either case, a freebody diagram should be drawn, and the force balance studied.

None of these devices should be used without initial calibration tests. Regardless of the claims made in advertising for commercially available devices, the characteristics must be known firsthand by the investigator. If the investigator cannot "burn rubber" personally on various surfaces to predict reasonably accurate results with the device, it should not be used. In such cases, the better approach may be to use tabulated values augmented by the investigator's own judgment of surface condition. The best technique available should be used to determine reasonable values for coefficients of friction.

6.2.3.b Inspecting the Vehicles

Information gathered from an inspection of the vehicles is used primarily in the analysis of postimpact and collision behavior. The items of general interest are brakes and steering, as the condition of these mechanisms may place limitations on the behavior of the vehicle after impact. A wheel that is pinned in a collision typically provides maximum braking at all times, regardless of driver braking effort or the record of skid marks.

Rear wheels often are not involved mechanically in a collision; however, rear driving wheels are seldom free to roll with negligible resistance. If an inspection of the engine compartment reveals that the engine likely quit running at impact, and if the vehicle is equipped with a manual transmission, then the driving wheels may have offered substantial resistance to rolling, regardless of the skid marks. If the transmission is a fluid-coupled device, less, but still substantial, resistance might be offered.

If all four tires left continuous skid marks, the foregoing sort of information loses its significance. If some of the marks are indistinct, at least for a portion of the trajectory, then the information is important. The latter is the norm in cases severe enough to warrant reconstruction effort, because in such cases, the driver may not be able to apply braking effort, or change steering input, after impact.

Portable scales provide an estimate of weight distribution on each wheel after impact. The total of these values should be a reasonable estimate of the weight of the vehicle (less occupants and contents) at the time of the accident. The individual values are not likely to represent a proper force distribution before impact, as postimpact values may be influenced by a twisted vehicle body or frame.

If portable scales are not available, the vehicle may be towed to public scales to obtain its total weight, or a reasonable "dry weight" may be obtained from registration records. When a vehicle is initially registered, the manufacturer's statement of origin (MSO) will list its manufactured weight, without fluids, and many states transfer this number to permanent records for tax purposes. Some states do not transfer the information when a vehicle's registration is transferred from another state, and instead insert another number that is one pound greater than the minimum for the applicable tax class.

If none of these techniques is fruitful, typical weights have been tabulated by vehicle type and wheelbase. These estimates may be used, with attendant range considerations.

If the tools for analysis include techniques to evaluate crash dynamics, measurements describing the deformation (or crush shapes) are needed. Typically, the dimensions needed relate to changes in body shape as measured from an original shape. Hence, it is often of benefit to be able to compare measurements of the damaged vehicle with a similar, undamaged vehicle.

Another caution is in order. The investigator must understand the model(s) on which the analysis techniques are to be based. It would not be unusual to find that a model is two-dimensional. The analyst is then left to determine appropriate ways to handle deformations at different elevations as seen in the plan view. Such deformations may arise when the bumper of one vehicle overrides or underrides that of another. The data should be complete enough to allow the situation to be analyzed in a reasonable way.

Inspections regarding mechanical function are beyond the scope of this book. The investigator should be familiar with the characteristics of braking and steering systems, at a minimum. It also is helpful to know the characteristics of tire and light damage. These and other mechanical factors may contribute to the cause of an accident, and an analysis of their appearance may help to explain the accident to the client, to an attorney, or to a jury.

6.3   ANALYTICAL RESOURCES, SKILLS, AND METHODS

6.3.1   The Drawing

The first step in an analysis is always preparation of a drawing. After that, the analyst may strike out in any one of several directions, in response to the unique character of the case. The drawing is prepared with a view toward the work to come, and toward a finished presentation. Format of the final presentation frequently suggests an appropriate scale; however, the analysis may require a different scale.

A formal report is usually typed on $8\frac{1}{2} \times 11$-inch paper and it may be placed in a binder that is slightly larger. If a drawing is to be most useful to readers of the report, it should be readily available while still conforming to the binder constraints. Therefore, the drawing should be reproducible so that it will fit neatly on an 11-inch-high page. It may be reproduced in any length that corresponds to the 11-inch height, as it may be fan-folded when the report is bound.

The analyst requires a scale that clearly illustrates the marks observed at the scene and still leaves room on the drawing board to position the vehicles at potential places of first sighting. This scale is likely to be 5 or 6 feet to the inch. If there is significantly more than a normal highway width placed vertically on the drawing, it will probably require reduction when it is reproduced for the report. The scale of the finished reproduction in the report should ideally be a standard scale, also. If the drawing is to be reduced for the report, a graphic scale should be provided, in addition to the standard notation.

The initial drawing should include all of the referenced features, but should not include vehicle outlines in the impact area. The drawing will be completed for presentation eventually, but only after all of the calculations have been completed. It is wise to use one or more copies of the initial drawing to work up position-related information. In this phase, thin cardboard scale models of the vehicles are useful tools. Using the workup drawings, the analyst determines the paths of critical features, such as an individual tire or the center of mass. In this way, distances that were not directly measurable at the scene can be scaled from the drawing. Computer-aided drafting (CAD) techniques are particularly helpful in this iterative process.

6.3.2 Distance, Speed, and Time

The second objective of the analysis is to provide as much insight as possible into the relationship of distance, speed, and time to the outcome of the incident. The direction of the analysis may seem indistinct in this phase of the project, but there is a pattern. There are several ways to approach the problem, depending on the tools available. Each approach endeavors to clarify the speeds and speed changes for each vehicle.

This information enables the analyst to compute the position of each vehicle at any given point in time. It also allows for the mathematical investigation of alternative scenarios; for example, "If vehicle number 1 had been traveling slower when its driver recognized danger, what would likely have happened if the driver chose to take the same action at the same position?" The variables that form the basis for such a study are shown graphically in Figures 6.2 and 6.3.

Figure 6.2(b) is the velocity triangle for a hypothetical vehicle number 1, in which V_{1b} is its velocity before impact, V_{1a} is its velocity after impact, and ΔV_1 is its velocity change as a result of impact. These three velocity vectors must form a closed triangle, by definition, just as the angular velocity "triangle" (or line) must close. Figure 6.2(c) is the same study for vehicle number 2. Figure 6.3 is a graphic statement of the conservation of linear momentum, showing the momentum components before and after impact. All of the momentum components are vectors also, and both of these two triangles must likewise close in common with the momentum vector for the system, G.

The objective is to be able to define all of these components. The typical procedure is to determine them one at a time, and then put them all together, so that any unknown quantity becomes evident. There will be instances when not enough information exists to close all of the triangles,

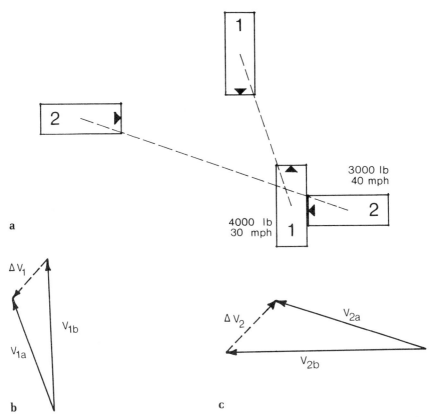

Figure 6.2 (a) Scene view of example collision. (b) Velocity triangle for vehicle 1 (V_1). (c) Velocity triangle for vehicle 2 (V_2).

and there will be other cases when there exist more data than are essential for a unique solution. All of the data should lead to solutions that cluster about a central theme.

The classical technique is to determine postimpact speeds using force and distance to define the energy dissipated in postimpact trajectories. These are then combined according to momentum concepts, and the speeds before impact result from knowledge of the preimpact directions. Another technique, which uses the computer, works with crush shapes to determine the ΔV values of impact. Combination of the ΔV values with speeds and directions before or after impact enables "closing" of the velocity triangles directly.

6.3.2.a Momentum

Most engineering dynamics textbooks contain a section on conservation of linear momentum. A review of this section will give important

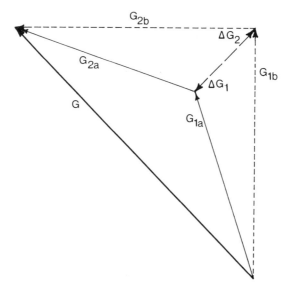

Figure 6.3 Momentum diagram for example shown in Figure 6.2.

momentum concepts for reconstruction. In a system of two colliding vehicles, where there are no forces external to the two vehicles, the linear momentum of the system (both vehicles collectively) does not change. If **G** is the symbol for the momentum of the system, Figure 6.3 shows that the vector (which has both magnitude and direction) is equal to the sum of the momentum vectors of the two vehicles.

Equation 6.1 states this relationship mathematically, where subscripts a, b, 1, and 2 refer to after impact, before impact, vehicle 1, and vehicle 2, respectively. The product of a vehicle's mass, m, and velocity vector, **V**, is its linear momentum, **G**, by definition. The masses are not subscripted with subscripts a and b, unless one breaks apart during impact.

$$\mathbf{G} = m_1\mathbf{V}_{1a} + m_2\mathbf{V}_{2a} = m_1\mathbf{V}_{1b} + m_2\mathbf{V}_{2b} \qquad (6.1)$$

Because **G** is the same before impact as after (barring any outside forces, such as interference with the pavement) the sum of the momentum vectors an instant before impact must be equal to the sum an instant after impact. The two masses are known and constant. The remaining four parameters are the before- and after-impact velocities of the two vehicles.

Note that the right-hand portion of Equation 6.1 is not the same as an algebraic expression relating four unknowns. There are eight properties involved in the relationship, both magnitude and direction for the four velocities. The expression can be solved only if two velocities are known, such as for both vehicles after impact, and if two directions are also known, such as for both vehicles before impact. The equation may be solved graphically by a diagram such as Figure 6.3 or numerically as shown in Table 6.1 later.

The drawing, in combination with scene data and vehicle inspection information, should be sufficient to establish the four directions. The only remaining values to be found are the four scalar values of speed for the two vehicles. The two that are most often desired are the speeds before impact, and they are obtained by first determining the speeds after impact.

6.3.2.b Work and Energy

To determine the velocities after impact, work-energy concepts are utilized. Any engineering dynamics text includes development of the equation

$$FS = \tfrac{1}{2} mV^2 \tag{6.2}$$

Equation 6.2 states that the work done, FS, on a mass m, which is the product of the force, F, applied in the direction of movement and the displacement, S, is equal to the *change* in kinetic energy of the mass, which is the entire right-hand term, simplified by the assumption that the vehicles come to rest from an after-impact velocity of V. Rearrangement of Equation 6.2 yields the left portion of Equation 6.3. Further substituting the vehicle's weight, W, divided by the accleration from gravity, g, for its mass, m, provides the right-hand term of the equation

$$V^2 = 2FS/m = 2(F/W)gS \tag{6.3}$$

The appropriate integral for the expression will be found in the text, in the section on kinematics, as the integral of the differential equation

$$v(dv) = a(ds) \tag{6.4}$$

That integral, assuming constant acceleration (a) is,

$$(V_{final})^2 = V^2 + 2aS \tag{6.5}$$

where V_{final} is the velocity after applying the acceleration, a (negative for deceleration), for the distance, S. Note the similarity between Equations 6.3 and 6.5; they are the same when V_{final} is zero and $(F/W)g$ is viewed as an acceleration factor. This is precisely the strategy used in reconstruction, except that F/W is given a special name, and the distance symbol, S, is replaced by d. The special name for F/W is "drag factor," and it is noted here as D_f. Hence, Equation 6.5 is rewritten as

$$V^2 = (V_{final})^2 + 2\,D_f\,gd \tag{6.6}$$

The distance traveled after impact and the mass for each vehicle are known quantities. If the force(s) acting on the vehicle as it slid to a stop were known, the speed of the vehicle at the beginning of the slide could be determined by using Equations 6.3, 6.5, and 6.6. If this can be done for both vehicles, the momentum expression can then be resolved to find the speeds before impact.

6.3.2.c Force, Mass, and Acceleration

To find the forces, force–mass–acceleration concepts are used. Aspects of vehicle center of mass and coefficients of friction must be studied to ultimately develop numerical values for forces, for the typical case. For the present, consider Figure 6.4, where it is assumed that the location of the center of mass is known, and that the coefficient of friction is the same for all tires throughout the final skid.

In the case of objects under acceleration, positive or negative, the summation of forces and moments is not equal to zero, as in the static case, but instead is related to the acceleration. The forces offered by braking of the wheels are shown on both axles in Figure 6.4, but their equivalent summation acting through the center of mass is not shown. In most texts, there is a phantom view showing that this resultant force, usually labeled R = ma, is to be considered in the dynamics. When a force is translated on a freebody diagram, there is an attendant couple. The mental image is confusing for some readers, but all have an appreciation for the "nose-down" effect that accompanies braking.

By application of the principles of dynamics, the following expressions can be developed:

$$\frac{N_1}{W} = \frac{B + h(a/g)}{A + B} \tag{6.7a}$$

and

$$\frac{N_2}{W} = \frac{A - h(a/g)}{A + B} \tag{6.7b}$$

Here, vehicle a/g may also be replaced by the equivalent D_f.

The total force acting to retard the vehicle in Figure 6.4 is the sum of the wheel-braking forces, $F_1 + F_2$. In considering the equation $F = ma$, it can be seen that all of the following is true.

$$D_f = F/W = \frac{F_1 + F_2}{W} = a/g \tag{6.8}$$

Equation 6.6 does not appear to refer to forces, because the notation used does not include any force symbols directly. Equation 6.8 illustrates that D_f contains this reference to forces in ratio form.

6.3.2.d The Simple Case: Maximum Braking

By definition, the upper limit of F_1 is the coefficient of friction, C_f, times the normal force, N_1. Similarly, the upper limit of F_2 is $C_f N_2$. Hence, if both wheels on both axles are known to be sliding for the full stop,

$$D_f = a/g = \frac{F_1 + F_2}{W} = \frac{C_f N_1 + C_f N_2}{W} \tag{6.9a}$$

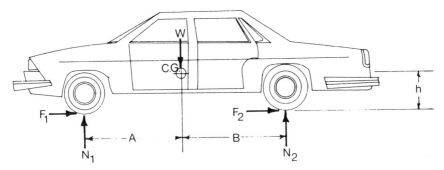

Figure 6.4 Center-of-gravity location and frictional forces resulting from braking.

or

$$D_f = \frac{C_f(N_1 + N_2)}{W} = C_f \qquad \textbf{(6.9b)}$$

After substitution into Equation 6.6, it becomes clear that, *if all four wheels on one vehicle leave skid marks from the point of impact to rest,*

$$V^2 = 2C_f g d \qquad \textbf{(6.10)}$$

where d is the distance traveled by the center of mass. If *both* vehicles leave four-wheel skid marks for the full distance, then both speeds after impact may be determined in this way, and the corresponding momentum vectors may be resolved. A convenient tabular form for developing the vector solution is found in Section 6.6.1. In a similar way, by use of Equation 6.6, four-wheel skid marks left before impact may be resolved to determine the speeds as brakes were applied.

Equation 6.6 is undoubtedly the most often used expression in traffic accident reconstruction. To the experienced forensic engineer, the relationships become second nature, in the same way that a pump engineer automatically thinks in terms of a head-capacity curve. The units V, d, and g must be compatible, usually feet and seconds, and the analyst should be able to convert between miles per hour and feet per second with ease. Engineers should also recognize that instantaneous coefficients of friction are not constant for the duration of a skid, as assumed in the integration leading to Equation 6.6; they are not even linear functions of speed.

Nevertheless, the dependence on the drag factor notation has created a mind set that is difficult to overcome. Average coefficients of friction have been developed with this notation in mind, and the accumulated testing effort to develop these coefficients alone is mind-boggling. There is nothing incorrect about Equation 6.6, or the concept that equivalent deceleration can be expressed appropriately as an average fraction or percentage of vehicle weight. However, if the analyst blindly follows this concept, it can lead to erroneous interpretations. The danger arises when one or more tires do not leave complete, continuous skid marks.

6.3.2.e *The Not-So-Simple Case: Less Than Full Braking*

When braking forces are not sufficient to lock all four wheels, the algebraic solution is complicated. For example, in a case where the front wheels provide no braking effort, the rear wheels provide maximum braking, $A = 48$, $B = 52$, and $h = 20$, Equation 6.7b reduces to

$$D_f = C_f (0.48 - 0.2D_f) \tag{6.11}$$

Equation 6.11 is difficult to solve, for the general case, as it will not separate. If $C_f = 0.65$, then $D_f = 0.28$ for this example. If all tires skid, the contribution from the rear tires is 0.23, which is less. The effect of any increase in stopping force is to decrease the normal force on the rear tires and increase the normal force on the front tires. This reaction is termed *weight shift*.

For partial braking problems, Equation 6.11 is most often solved by iterative computer techniques. Algebraic solutions may be developed with dimensions expressed in ratio form, such as in fractions of wheelbase, for example. This approach simplifies the appearance of the algebra. The investigator should derive these expressions for his or her own use. This is the best way to develop an understanding of the models and assumptions, and therefore the best way to ensure that a formula is used appropriately; it is also a good way to learn.

6.3.2.f *Center of Mass*

In the previous discussion of Figure 6.4, the center of mass (C.G.) was assumed to be located approximately midway between the wheels. This assumption is reasonably accurate, even for an empty pickup truck. If w_b represents the wheelbase, the expected distance from the C.G. to the front axle is $0.48w_b$, which implies that 52% of the weight is distributed to the front axle. An empty, long-wheelbase pickup truck weight may more likely be distributed 53:47 or 54:46.

If the vehicle in question is available for inspection, it is usually convenient to measure the weight on the front and rear axles. The front axle-to-C.G. fraction of the wheelbase is the ratio of the *rear* axle weight to total weight, and vice versa. Equation 6.7 will provide the necessary mathematics after a is set equal to zero. It is common to find the average driver's hip joint located close to the C.G. This point is known as the "H-point" in vehicle design, and the coincidence helps to provide a comfortable ride for front-seat passengers.

The elevation above the ground of the C.G. may also be estimated by averaging the height of the windowsill and lower body sheet metal. In truck tractors, the height of the C.G. is usually near the top rail of the frame.

In empty van trailers, it is just below the floor. If it is important to determine the location more accurately, the vehicle may be weighed with one axle lifted high, on the order of 30% of the wheelbase, with a high-boom wrecker, so that the only weight on the scale is one axle. "Bob" truck scales are useful for this weight. Because they are short, there is less tendency for them to bind when loaded on one end. The best way to measure the angle is by clinometer. An expression for the height of the C.G. is

$$h = \frac{G(w_b)}{W(\tan \alpha)} + R$$

(6.12)

where

G = gain in axle weight ($W_{elevated} - W_{level}$)
w_b = wheelbase
W = vehicle total weight
α = angle with the ground
R = radius of the loaded tire

When this expression is developed for a specific case, the effect of vertical movement of the axles should be considered. The scale mechanism should be observed for hysteresis and significant increment.

6.3.2.g Coefficient of Friction

Coefficients of friction are important variables in the study of a vehicle's behavior in an accident. The difficulty of determining appropriate values with precision was discussed in Section 6.2.3. There will always be some uncertainty about the values used. This is a fact with which the legal system has already grappled, and the techniques used for analysis should take the potential variance into account. The proper way to deal with the uncertainty is to establish a reasonable range for C_f, or for any other variable in question, and to explore the influence of this range on the end results.

Much has been written on the subject of friction. Many papers are available on tire friction, and there are many publications that summarize these reports. Some resources include a thorough discussion on the exponential character of an "instantaneous coefficient of friction" as a function of speed, and ways to predict an "average coefficient of friction" that can be related to D_f, the drag factor. Unfortunately for the reconstructionist (but fortunately for the motorist), the bulk of this research has been directed toward improving the wet coefficients of friction of tires on pavement. Dry sliding coefficients are principally a function of road surface characteristics. There is not much a tire designer can do about dry sliding coefficients and still provide reasonable tire life.

There are four coefficients of friction for any two contact surfaces:

wet sliding, wet static, dry sliding, and dry static. The static values of C_f are potentially important for some cases, but the principal interest in traffic accident reconstruction concerns wet or dry sliding coefficients. Published tables often provide a range of values for more than one type of C_f for various conditions and surfaces. The ranges include an allowance for scatter in the data on which they are based and for various surface conditions. The analyst can do very little about statistical scatter, but allowances can be narrowed for surface condition.

Fresh, dry asphalt, for example, initially offers a C_f of 0.8+ for panic stops of cars and pickup trucks from speeds on the order of 30 mph. This C_f should typically decay rapidly by 20 to 25% with the passing of traffic, however, and it should remain at that level for a long period of time. It should be reduced by 30% only after the surface becomes traffic polished, and possibly by more only after "bleeding." Concrete offers a similar resistance to sliding. These characteristics are relatively consistent for passenger car tires because most behave like 65 Durometer SBR (styrene–Butadiene–rubber), and most are loaded in the same manner and to the same degree.

Truck tires are different, however. Their tread compounds are different, and typical footprint stresses are three times that of automobile tires. They are also subject to much more variation in loading, because of the effects of cargo and cargo distribution. A review of the literature suggests that stopping distances for trucks are 15 to 25% greater than for passenger cars, and that speed estimates for trucks based on skid marks are more difficult to predict. The difficulty encountered is more complex than just the additional number of tires that must be considered.

If a broad range for D_f is used, the results may predict speeds that have too much range to be of practical use. On the other hand, predicted speed is a square root function of the product of D_f and distance (Equation 6.6). Such estimates may be satisfactory, as the effect of a potential error in measurement is diminished through the mathematics. If it is reasonable to assume that the estimate of the product is within 6% of the true value, than the resulting estimated speed will be within 3% of the true value. Knowing this character of the physics and the objective of the work, the analyst should be in a position to determine the degree of effort required to obtain appropriate accuracy in estimated values for the variables.

6.3.3 Collision Dynamics

The basic physical parameters that are important in traffic accident reconstruction have now been presented. The theory is relatively straightforward for those with engineering backgrounds, but the mathematical detail quickly becomes complicated, involving a large number of variables. It is not surprising, then, that the computer is often used in the analysis to keep the large number of relationships between variables in proper perspective.

6.3.3.a Trajectory

The application of computers to traffic accident reconstruction technology has led to the practical use of concepts that would be unthinkable to use by hand. For example, it is now possible for the analyst to look at each individual tire print at any point in time, and sum all of the forces on all of the tires, taking into account whether each individual tire was rolling, sliding, or yawing. The effect of these forces, acting over a short period of time, in changing angular and linear velocities can be determined. Then the same can be done for the next small period of time. The time required to perform the calculations by hand would be prohibitive, but a computer can run a complete trajectory in less time than it takes to study the computed results.

There are several commercially available software programs for trajectory or vehicle-handling studies. Some must be run on large-capacity computers; others can be used on desk-top machines. Some programs stand alone; others are used in conjunction with impact analysis programs. Each program contains a model of a vehicle as the fulcrum on which it is balanced. It is essential for the user to understand the model, and assumptions for the parameters, prior to relying on results obtained from the program.

6.3.3.b Impact Analysis

Routine postimpact speed estimates are easily determined by hand calculations. Given the required data, meaningful estimates of speeds before impact can also be found easily, using the linear momentum concepts discussed previously. With these results, distance, time, and speed relationships can be developed. One parameter, however, is very difficult to determine by hand calculations: the angular velocity. Also, it is difficult to know how to use this parameter in the analysis, even when it becomes available. This is the subject of software developed for impact analysis.

Impact analysis and simulation programs, with appealing names like *CRASH, CRUNCH, WRECKER,* and *SMAC,* typically provide changes in both linear and angular velocity that are expected in response to changes in body shape. If the velocity after impact is known and the change of velocity during impact is known, all that is required to obtain the velocity before impact is to close the velocity triangle.

The impact ΔV appears to be a totally new piece of information at first, but it is found by impulse-momentum considerations. This concept is not totally independent of the more familiar linear momentum concept. However, it is based on a new and different source of information, the energy of crush shapes. This makes the impact ΔV an important variable in the reconstruction process. If the ΔV and classical momentum solutions do not agree, the investigation should be expanded to determine why. The two solutions should agree.

The new datum is the change in angular velocity. With this information, a new independent vector system is available, and it must close, just as the linear velocity triangles must close. There will be times when the data are not complete enough to use this approach. When the data are available, however, it may help to establish a narrower range for estimated variables, such as C_f.

6.3.4 Analysis of Variance

Once the most likely scenario for a traffic accident has been determined, it is proper procedure to investigate the effects of potential variances in the data. Good practice includes studies of such variances routinely, as the solutions are developed. The values used for C_f are obvious targets for this exercise; others may not be so obvious at the outset. For example, contact angles [also known as principal directions of force (PDOF)] may not be known with sufficient precision, and these normally have a great influence on changes in angular velocity. Furthermore, variances in contact angles may relate to variances in position, particularly the position of a center of gravity, at separation. Such variances in position are usually coincident with small changes in separation velocity. The analyst must be reasonably certain that the variance in results will not affect a judgment regarding any critical issue. Otherwise, the stated judgments must be broad enough to properly reflect the uncertainty involved in the variances.

6.3.5 Sequence of Events

The work is not complete when the analysis produces an explanation of what happened in the particular case under investigation. In the typical case, further studies should be made to see what the outcome of the affair would have been in different circumstances. The details of the case will suggest alternative circumstances to consider.

For example, consider the case of a motorist who stops at a stop sign, then proceeds into the intersection after looking both ways. Unfortunately, the driver misjudges the speed of a car coming from the left, which is traveling 5 to 10 mph faster than the posted speed limit. In the analysis, it becomes evident where the driver of the speeding car first recognized the danger; it is at a distance corresponding to the driver's reaction time before the skid marks are first observed. The analyst can substitute the posted speed limit, calculate the distance required to coast from that point for the same reaction time, and then find the time required to skid the greater distance to the point of impact. In the meantime, the first driver may have been able to drive through the intersection. It would not be unusual to find that a reduction in speed of 5 mph would have resulted in a miss, albeit a near collision, rather than simply a collision of lesser intensity. This type of information helps to put the details of the incident in proper perspective for all interested parties.

6.4 THE REPORT

6.4.1 Formal Reports

Formal reports of traffic accident investigations are much the same as reports for any other forensic endeavor. Most begin with a synopsis of background information and continue with observations gathered at the scene and through inspection of the vehicles. After this, the calculations are presented and discussed. Normally, a summary or conclusion is provided, after an interpretation of the relevant information. Captioned, mounted photographs typically follow the text. A summary of selected calculations and other relevant material, such as a copy of the police report, are usually included in an appendix.

6.4.2 Other Report Types

There are at least three different ways to present the results of an investigation to a client: (1) an informal conference, (2) a brief letter, and (3) a formal report. Occasionally, an informal conference is followed by a brief letter. In each case, the desires of the client dictate the format of the report. Regardless of the format selected, material should be retained in the file to support any conclusion expressed in the report.

6.5 BEYOND THE REPORT: EXAMPLES OF INCREASED SAFETY AS A RESULT OF FORENSIC INVESTIGATIONS

An engineer who does thorough work and explains it well may never hear about the matter after the report has been transmitted. The report is likely to be circulated to the affected parties, who may reach some kind of agreement among themselves. This is a case where the system works well. Once in awhile, however, an engineer finds that the work has had an influence beyond the case it was intended to serve. Sometimes the engineer personally chooses to take it beyond the case. Prior to going beyond the case, the engineer should consult with the client to be sure there are no objections.

6.5.1 Tom Prewitt's Work: Perception of Sirens

One example is the work done by Tom Prewitt on the perception of sirens. It is now understood that the typical motorist does not hear sirens as well as once thought. Tom Prewitt prepared, for presentation to emergency vehicle drivers, a videotape explaining the situation. Many cities now have the videotape in their libraries. This work was reviewed in the February 1986 issue of *Mechanical Engineering* magazine.

6.5.2 Front-Axle Brakes on "Bob" Tractors

Another example of the influence of forensic work on safer practices affects truckers who earlier followed federal regulations in removing tractor front-axle brakes. This practice was encouraged to improve performance on icy roads, so that the driver had better steering control. The obvious trade-off was in stopping ability. At first glance, the trade-off appears to be acceptable. The problem, however, occurs when a trucker drives such a tractor without a trailer.

This practice is called "running bobtail" or "bob," for short. A "bob" tractor without front-axle brakes cannot comply with federal braking guidelines. There simply is not enough weight on the rear axle(s) to provide sufficient stopping force. One response to this situation has been to inform the drivers and maintenance personnel in trucking organizations that "running bob" with no front-axle brakes is an unusually dangerous practice. The federal guidelines that promoted the practice have now been rescinded.

6.5.3 Suspension Bolt Recall

One other example involves an automobile that was reported to have been traveling on a two-lane road at a moderate speed when it suddenly veered to the right and overturned. A rear suspension bolt had given way and allowed a trailing arm to swing free. This allowed the rear axle to move back far enough to let the front of the drive shaft drop to the pavement. Some of the bolt was found up inside the bracket, in the frame. The forensic investigator's report explained the result of a metallurgical examination of the bolt. Several months later, the manufacturer recalled that series of automobile for replacement of the bolts.

6.6 CASE STUDIES

6.6.1 Rear-End Case Study

The first case selected for illustration concerns a small pickup truck, vehicle 1, which struck a full-size pickup, vehicle 2, from the rear on an interstate highway. In addition to the general character of vehicle behavior, the principal questions related to the speed of vehicle 1 and the origin of impact. The available data included a tire mark left on the white shoulder stripe (recorded in photographs) and a plowed furrow made by one of the wheels of vehicle 2, which was still evident 8 months after the incident. A scene drawing is provided as Figure 6.5.

This case is of interest for several reasons. It illustrates a technique for locating the origin of impact, and it shows how crush, skid,

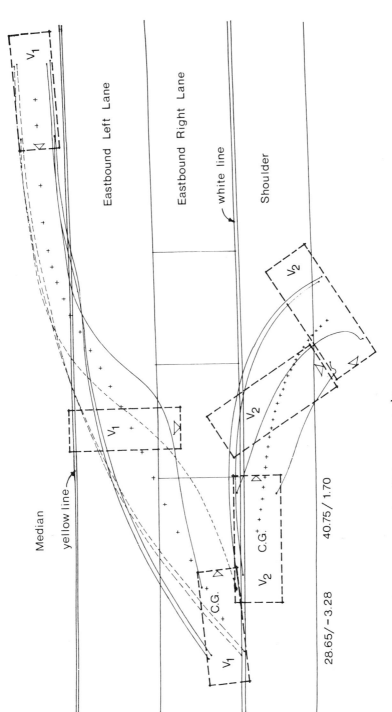

Figure 6.5 Scene drawing: rear-end collision case study.

and linear momentum data can be combined to achieve an understanding of the essential character of a mishap. The work of determining the origin is based on a computer software analysis of a mathematical model of the vehicles.

One segment of the software (*TRAJECTORY*) determines the forward, lateral, and angular velocities required to cause the modeled vehicle to follow a prescribed path after impact, with each vehicle analyzed separately. Another segment (*CRASH*) determines the change in these parameters that results from a collision causing the observed crush shapes, with both vehicles analyzed simultaneously. The procedure is to determine the data for vehicles presumed to start at alternative impact positions, located on a grid, but always in matching relative positions. Trends will develop in the data, as they are analyzed in sets, which are organized by starting grid position. The data tend to converge at the approximate position where impact occurred.

In this case, the driver of vehicle 1 stated that he steered hard left in one motion (tire steering between 3° and 5°) without applying brakes before impact, and that impact occurred quickly thereafter. A simulation of this condition shows that the vehicle's rotation, or angular velocity, builds in steps, or plateaus to a maximum value (Chi et al. 1984). The result of the simulation is that the preimpact rotation of vehicle 2 should be expected to be between zero and -0.45 rad/sec. As vehicle 2 was assumed to be initially stationary, its preimpact velocities were assumed to be zero.

One of the physical tools available for analysis is the closing of velocity triangles, as discussed earlier. Angular velocities must also close. A vehicle's rotation after impact must be equal to its rotation before impact plus the change during impact (Figure 6.2). The collision analyses provide estimates of the change in rotation caused by impact as a function of the location of impact. Trajectory data provide independent estimates for the rotation after impact needed for the vehicle to follow the path observed.

When these data are combined graphically, they are seen to converge in a relatively small area of impact location. An example is provided in Figure 6.6, where postimpact rotation of vehicle 2 along a common ordinate is shown varying along the abscissa. The effect of variance of the postimpact rotation of vehicle 2, with respect to the preimpact rotation of the other vehicle, is due to small changes in the principal direction of force thereby created. Many such graphs could be shown, one for each selection of common ordinate. A better illustration might be a three-dimensional surface plot, which is not shown. The area in which the data converge well is a relatively narrow and surprisingly short strip of highway real estate that runs at an angle to the roadway. It converges to place the portion of vehicle 2 that was struck at least partially in the roadway.

Vehicle 2 was assumed to have been parked when struck. Hence, the ΔV predicted by *CRASH* could be compared directly to the required velocity predicted by *TRAJECTORY*. These straight lines crossed at a starting abscissa near 40.25, the same abscissa that is predicted for the

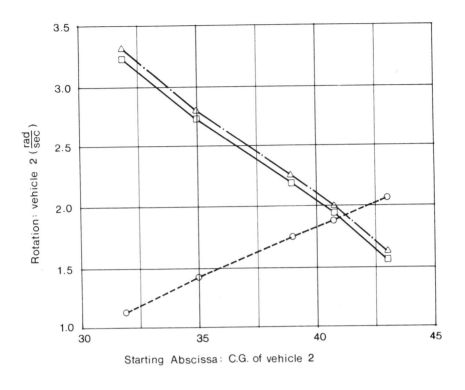

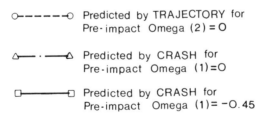

Figure 6.6 Vehicle 2 postimpact angular velocity versus abscissa location at impact.

velocity triangle of vehicle 2 to close without any preimpact linear velocity. Other parameters may be compared for fit with the observed data, which vary from case to case. Exploring this example in the vicinity of the x = 41, y = 1.8 location led to the recognition that the most likely fit, with all of the observed data taken into account, is the point x = 40.75, y = 1.7.

In addition to the initial location of vehicle 2, the initial speed of vehicle 1 was of interest. The value of the speed predicted by those analyses that were reasonably close to the located origin all clustered around 60 mph. The raw datum having the greatest potential for variance was the coefficient of friction. This value was estimated solely on the basis of experience, in combination with standard published tables. An analysis of variance of the results, with respect to this parameter, showed that if the value used for C_f was 0.75 instead of 0.65, as initially assumed, the predicted speed would increase from 60 to 62.4 mph. This small change in the outcome may seem surprising, but it can be understood when Equation 6.6 is carefully reviewed.

The classical energy/momentum solution requires the separation speeds to be estimated from the energy absorbed by the tires as the vehicles come to rest. The two momentum vectors are then added to determine the total momentum of the system. The sum of the two momentum vectors before impact must equal their sum after impact. As the preimpact momentum of vehicle 2 in the example was assumed to be zero, the other vehicle, vehicle 1, must have accounted for all of the preimpact momentum. Hence, the speed of vehicle 1 is determined to be approximately 60.5 mph. In Table 6.1, velocities after impact were estimated using the energy expression $V^2 = 2D_f gS$, where subscript 1 refers to vehicle 1, subscript 2 refers to vehicle 2, and $D_f g$ is a composite average deceleration based on an estimate of the portions of travel forward, sideways (where $D_f = C_f$), and backward. D_f includes the effect of weight shift, as both vehicles were braked principally by their rear wheels. In the energy expression, S is the distance traveled after impact (assumed to be the straight line connecting the centers of gravity between positions of impact and rest).

If the values for velocity after impact found by the *TRAJECTORY* computer analysis are substituted for the values in Table 6.1, the momentum solution predicts a preimpact speed for vehicle 1 of 59.1 mph. If the velocity triangle for vehicle 1 is closed using *TRAJECTORY* velocities and *CRASH* ΔV values, a preimpact velocity for vehicle 1 of 58.3 mph is predicted. The latter approach also predicts an initial velocity for vehicle 2 of −0.2 mph. All of these estimates for the initial velocity of vehicle 1 should be equivalent.

The results of the study are shown in Figure 6.5. The east–west x coordinates are arbitrary, but increasing to the east (right). The north–south y origin is the south (bottom) edge of the white stripe at the tire mark. The roadway at the scene is a 2546-foot-radius curve, which has been superimposed over the two trajectories.

Note that the starting position of vehicle 1 is engaged with the other vehicle by approximately 5 feet, where most studies would place the two vehicles bumper-to-bumper. A separate drawing was prepared from data generated using a 0.01-second time interval. With starting positions bumper-to-bumper, the two separated without engaging. With starting positions as shown, the implied match with vehicle damage was uncanny. We have no reason to expect a close representation of data in the impact area, however, as no effort was made to simulate detailed crush behavior. Impact forces are averaged in the referenced *CRASH* software.

Vehicle 1 is shown starting at a proper angle *relative* to the vehicle struck. The study revealed that the angle of vehicle 2 to the

Table 6.1 Energy/Momentum Solution for Rear-End Case Study

For vehicle 1: $D_{f1} = 0.512$, $S_{1x} = 70.35$, $S_{1y} = -24.79$
For vehicle 2: $D_{f2} = 0.494$, $S_{2x} = 28.25$, $S_{2y} = 10.10$

	Angle	Weight × Velocity		gMV		gMV_x	gMV_y
AFTER IMPACT							
G_{1a} (vehicle 1)	−19.41	2851 × 33.737	=	96,183	=	90,716	−31,966
G_{2a} (vehicle 2)	19.67	4138 × 21.016	=	86,966	=	81,890	29,277
System: **G**	−0.89	Total:		172,626		172,605	−2,689
BEFORE IMPACT							
G_{1b} (vehicle 1)	−0.89	2851 × 60.549	=	172,626	=	172,605	−2,689
G_{2b} (vehicle 2)	NA	4138 × 0.000	=	0	=	0	0

roadway before impact was of no material significance in determining the origin. Hence, the position is shown parallel to the roadway, and all of the relative shift is shown in the angle of the small truck, vehicle 1.

6.6.2   Right-Angle Case Study

This example case involves a vehicle, vehicle 1, that collided with the side of another vehicle, vehicle 2. It is of interest because it illustrates how a project can change direction in midcourse. This example also illustrates the importance of studying the sequence of events and piecing together various witness statements.

Photographs were taken at the scene before the vehicles were removed. These photographs helped to locate the impact and rest positions of the vehicles, as the investigating officers did not record distances in their report. It might be of interest to note that the rest positions were located partially by studying photographs taken at right angles. These photographs showed the positions of the vehicles relative to identifiable permanent features of the landscape. The positions were recreated at the scene by viewing the features from the photographer's position, and recording intercepts of line of sight with curb lines and other features. These dimensions were transferred to a drawing, so that they formed intersecting lines, thereby locating the vehicles on the drawing.

The photographs also identified the path followed by the front wheels of vehicle 1, which left heavy skid marks as it approached impact. Data were available to obtain reasonable estimates for the postimpact speed of vehicle 1 and its change in speed as a result of impact. Analysis of the preimpact skid distance, in combination with a "full-braking" scenario, resulted in a speed estimate greater than that given by witnesses.

A subsequent reevaluation of the data, bringing into question the braking effort of the rear wheels, suggested that the rear wheels might not have contributed any braking effort at all. Hence, a mechanic was employed to disassemble the rear brakes. The rear brakes were found incapable of contributing any meaningful stopping force prior to impact.

Whether or not impact could have been avoided altogether with adequate brakes was not possible to determine in this case. However, the factors generally associated with personal injury were significantly more severe as the result of inoperable rear brakes than they would have been with serviceable brakes. Determining the magnitude of these factors is about as far as an engineer can proceed with the data, without access to medical expertise.

References

Anderson, J. 1985. Limitations of the AASHTO formula for braking distance. *Journal of the National Academy of Forensic Engineers* 2, No. 2 (December): 33–44.

Baker, J. S. 1975, *Traffic Accident Investigation Manual*. Evanston, IL: The Traffic Institute, Northwestern University.

Baumeister, T. ed. 1967. *Marks' Standard Handbook for Mechanical Engineers*. New York: McGraw–Hill, NY, 7th ed.

Brach, R. M. 1987. A review of impact models for vehicle collision." Paper presented at the SAE Congress, Warrendale, PA, February 1987, SAE Paper No. 870048.

Chi, M., and J. Vossoughi. 1984. Forensic engineering for automobile accident reconstruction. *Journal of the National Academy of Forensic Engineers* 1, No. 2 (October): 39–56.

Flick, R. 1987. Benefits of front brakes on heavy trucks. Paper presented at the SAE Congress, Warrendale, PA, February 1987, SAE Paper No. 870493.

Fonda, A. 1987. *CRASH* extended for desk and hand-held computers. Paper presented at the SAE Congress, Warrendale, PA, February 1987, SAE Paper No. 870044.

Gardner, J. D. ed. 1982. *Highway Truck Collision Analysis*. New York: American Society of Mechanical Engineers.

Hicks, J., and R. Hess. 1986. The application of computer programs in forensic engineering. *Journal of the National Academy of Forensic Engineers* 3, No. 1 (June): 17–48.

Manning, L., and L. Bentson. 1984. Highway speed vs. sideslip (Critical speed in a curve). *Journal of the National Academy of Forensic Engineers* 1, No. 2 (October): 3–17.

_____ 1985. Enhancing CAD for the forensic engineer. *Journal of the National Academy of Forensic Engineers* 2, No. 1 (April): 33–44.

Meyer, W. E., et al. eds. 1981. *Frictional Interaction of Tire and Pavement*, ASTM (PCN) 04-793000-37. Philadelphia: American Society for Testing Materials.

Moffat, C. A., et al. eds. 1980. *Highway Collision Reconstruction*. New York: American Society of Mechanical Engineers.

National Automotive History Collection. Technical Data Collection. Detroit: Detroit Public Library.

National Highway Traffic Safety Administration (NHTSA). 1985a. *NHTSA Heavy Duty Vehicle Brake Research Program Report No. 1. Stopping Capability of Air Braked Vehicles, Vol. I: Technical Report*, PB86-108628. East Liberty, OH: NHTSA, April.

_____ 1985b. *NHTSA Heavy Duty Vehicle Brake Research Program Report No. 4. Stopping Capability of Hydraulically Braked Vehicles, Vol. I. Technical Report*. PB86-149267. East Liberty, OH: NHTSA, October.

Radlinski, R. 1987. Braking performance of heavy U.S. vehicles. Paper presented at SAE Congress, Warrendale, PA, February 1987, SAE Paper No. 870492.

Rogers, K. B. 1981. *Methods of Recording and Using Accident Information*. Little Rock, AK: Arkansas State Police.

Society of Automotive Engineers (SAE). 1987a. *Accident Reconstruction: Automobiles, Tractors Semitrailers, Motorcycles and Pedestrians*. A collection of papers presented at the SAE Congress, Warrendale, PA, February 1987, SAE No. P-193.

_____ 1987b. *Vehicle Highway Infrastructure: Safety Compatibility*. A collection of papers presented at the SAE Congress, Warrendale, PA, February 1987, SAE No. P-194.

Schreier, H. 1987. Applicability of the EES-accident reconstruction method with Mac-CAR. Paper presented at the SAE Congress, Warrendale, PA, February 1987, SAE Paper No. 870047.

Struble, D. 1987. Generalizing *CRASH3* for reconstructing specific accidents. Paper presented at the SAE Congress, Warrendale, PA, February 1987, SAE Paper No. 870041.

Williams, R. 1982. *Legal Aspects of Skidmarks in Traffic Cases*, 5th ed. Evanston, IL: The Traffic Institute, Northwestern University.

Woolley, R. 1987. The *IMPAC* program for collision analysis. Paper presented at the SAE Congress, Warrendale, PA, February 1987, SAE Paper No. 870046.

7. Transportation Disaster Investigation

RUDOLF KAPUSTIN

7.1 INTRODUCTION

Forensic engineering for transportation accidents includes such diverse investigations as aviation disasters, railroad derailments, and trucking accidents involving transportation of hazardous materials. The overriding concern in such investigations is prevention of future accidents. As a result of formal investigations, products are refined and operating procedures are improved so that recurring accidents involving deficient products or ineffective procedures are minimized.

My own experience has been primarily in the field of aviation accident investigation. This chapter emphasizes aviation accidents, but procedures of investigation for other transportation accidents are similar.

A great deal of effort has gone into establishing the procedures currently used in aviation accident investigation. This discussion outlines the coordinated team approach to investigation. A case study, the Air Florida Potomac River accident of January 1982, is used to illustrate investigation concepts.

Refined procedures for investigating accidents and disseminating information resulting from the investigations have great potential for reducing the frequency and severity of future accidents in the transportation industry.

7.2 HISTORY

7.2.1 First Fatal Aircraft Accident

On the afternoon of September 17, 1908, the first fatal accident involving a powered aircraft occurred. Fort Meyer, Virginia, on the outskirts of Washington, DC, was the location. U.S. Army Cavalry Officer Lieutenant Thomas E. Selfridge was the unfortunate first person to be killed as a result of the technology of powered flight.

With Orville Wright at the controls, the aircraft *Flyer 3* was scheduled to make a demonstration flight. Lt. Selfridge, a member of the Army's acceptance board, was on board as a passenger to demonstrate to the U.S. Army that powered flight was a reality.

The *Flyer* had been retrofitted that morning with a larger diameter propeller. The takeoff was routine, and the airplane had made several passes over the demonstration area when disaster struck. First, a violent vibration in the propeller and then secondary damage to a guy wire and rudder control system forced Orville Wright to attempt an emergency landing. Substantial loss of lift and stability resulted in an uncontrolled descent and a catastrophic crash. Orville Wright survived his injuries; Lt. Selfridge did not.

The first known aircraft accident investigation followed this incident. The cause of the accident was determined with state-of-the-art precision. The longitudinal separation of the propeller and the ensuing failure sequence was documented. This investigation was probably the first phase in the development of transportation accident investigation technology.

7.2.2 Need for an Aviation Accident Authority

With the development of more advanced aircraft and transportation systems, along with the development of investigation methodology, came the formation of official civilian and military accident investigation authorities. Catastrophic transportation accidents have major social and economic implications. Because of this, the investigation of such accidents was often conducted in a hurried atmosphere, sometimes in a haphazard manner. Often the investigations were dominated by intense political or economic pressures.

The need for investigative bodies with complete independence from other governmental regulatory agencies or civilian special interest groups became evident. If unbiased, objective investigations were to contribute to accident prevention, such independence was clearly a necessity.

It was soon found that an agency charged with the responsibility of accident investigation could not in most cases remain totally objective, if the same agency had regulatory responsibility over the transportation mode under investigation. Because such an agency may have been responsible for creating or approving a regulation or procedure involved in the accident, there was a strong tendency to exonerate, cover up, or dismiss the effect of such a procedure in contributing to the accident. Likewise, if the agency were involved in pilot or operator certification, there could be reluctance to find fault with the certification or licensing process. If the agency inspected the vehicle involved in the accident and a mechanical defect contributed to the accident, the agency might be reluctant to admit negligence in inspection procedures.

Today it would not be expected that the Federal Aviation Administration would conduct an objective, unbiased investigation of its own air traffic control system, if an aviation accident deeply involved air traffic control procedures. Establishment of the independent U.S. National Transportation Safety Board in 1974 was the outgrowth of these concerns.

7.2.3   The National Transportation Safety Board

The initial groundwork for the eventual establishment of the National Transportation Safety Board can be traced to the Air Commerce Act of 1926. This legislation provided the Department of Commerce with the authority to determine the cause of aviation accidents. The actual investigative work and cause determination were done by the Aeronautics Branch, a small unit within the Department of Commerce. The Aeronautics Branch had, by 1933, evolved into the Bureau of Air Commerce. Sometime later, the Bureau of Air Commerce was replaced by an "independent" Air Safety Board. In 1940, the U.S. Congress created the Civil Aeronautics Board (CAB), whose statutory functions included the investigations of aviation accidents for the purposes of determining their "probable cause" and making appropriate safety recommendations to prevent the recurrence of accidents. The unit within the CAB responsible for the aviation safety function was the Bureau of Safety. Other bureaus within the CAB were responsible for economic regulation of air commerce and the enforcement of economic regulations.

A new cabinet-level Department of Transportation to oversee all modes of U.S. transportation was created, and the entire Bureau of Safety and its aviation safety function were moved, as a semi-independent entity, into the Department of Transportation in 1966.[1] The new entity was named the National Transportation Safety Board. The five members of the NTSB were appointed by the President of the United States. In addition to the aviation safety function, the NTSB was given the additional responsibilities of investigating highway, railroad, pipeline, and major marine accidents.

The regulatory agencies for the various modes of transportation,

---

[1] The National Transportation Safety Board was established by statute in 1966 (Public Law 89-670; 80 Stat. 935) as an independent agency, located within the Department of Transportation.

such as the U.S. Coast Guard for marine activities and the Federal Aviation Administration for aviation, were directly responsible to the Secretary of Transportation under the 1966 arrangement. The U.S. Congress, seeking complete independence for the NTSB, passed the Independent Safety Board Act of 1974.[2] This act gave the NTSB more authority in transportation accident investigation, and gave the Board complete administrative independence from the Department of Transportation and its regulatory and operating agencies.

7.3 ACCIDENT INVESTIGATION METHODOLOGY

7.3.1 U.S. Experience in Accident Investigation

Before creation of the NTSB in 1966, aviation accident investigation was, for all practical purposes, the only formalized transportation accident investigation activity in existence in the United States. Likewise, other countries with major air transportation systems had, over the years, developed accident investigation methodologies. International standards for the format and general content of an aviation accident report were formulated by the International Civil Aviation Organization (ICAO)[3] in 1966. The United States played a principal role in establishing the report format; in fact, the basic format had been in use in the United States prior to its adoption by ICAO as the international standard. Although the report format did not directly dictate the structure of the accident investigation organization, it was suggestive in encouraging the "team" concept employed by the U.S. Civil Aeronautics Board Bureau of Safety, and later the NTSB aviation accident investigation unit. This unit is currently known as the Aviation Accident Division of the Bureau of Accident Investigation. (See Figure 7.1.)

---

[2] Title III Sect. 302 of the Independent Safety Board Act of 1974 states in part the findings of Congress:

> Proper conduct of the responsibilities assigned to this Board requires vigorous investigation of accidents involving transportation modes regulated by other agencies of Government; demands continual review, appraisal and assessment of the operating practices and regulations of all such agencies; and calls for the making of conclusions and recommendations that may be critical of or adverse to any such agency or its officials. No federal agency can properly perform such functions unless it is totally separate and independent from any other department, bureau, commission or agency of the United States.

[3] In 1944, delegates from 54 countries attended a civil aviation conference in Chicago to draft multilateral agreements on international air operations and a convention establishing the International Civil Aviation Organization (ICAO). ICAO reached an agreement with the United Nations on December 14, 1946, and became a functioning organization on April 4, 1947. The organization operates under the auspices of the United Nations with headquarters in Montreal, Canada. Its main function is the development of international air transport and techniques of air navigation.

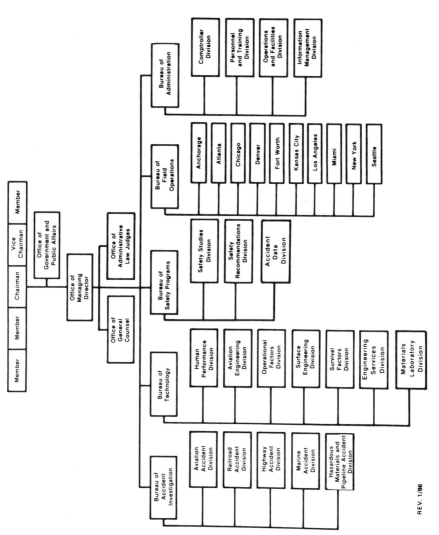

Figure 7.1 Organization of National Transportation Safety Board.

NTSB staff personnel, the nucleus of the newly added transportation accident investigation modes when the independent Safety Board was created, were highly qualified in technical expertise relative to specialized transportation modes. There was, however, relatively little knowledge of advanced accident investigation methodology, such as was being applied by the aviation function of the NTSB. Consequently, the aviation mode, with its extensive experience in accident investigation methodology, became a role model for the other newly formed groups.

7.3.2 The Investigation "Team" and "Party" Concept

7.3.2.a *The Concept—How It Works*

The legislative authority pertaining to accident investigation for the other transportation modes was not quite as well defined as it was for the aviation mode. It was soon determined that the "team" and "party" system, used by the aviation investigators, was an effective means of organizing and structuring the investigation of any major or even noncatastrophic transportation accident (Figure 7.2).

The team, which in some cases is composed of over 100 persons, is always headed by a NTSB investigator-in-charge. Each of the approximately 12 disciplines, or technical specialty groups, are headed by a senior NTSB technical specialist, referred to as the group chairman. The "party" system is used because of the impossible task of keeping all NTSB technical specialists fully qualified in the operational and engineering aspects of each type of vehicle, aircraft, or vessel that may become involved in an accident.

In the case of an aviation accident, the NTSB investigator-in-charge may "designate parties to participate in the field investigation. . . . Parties to the field investigation shall be limited to those persons, government agencies, companies and associations whose employees, functions, activities or products were involved in the accident or incident and who can provide suitable qualified technical personnel to actively assist in the field investigation."[4]

The same rule that permits party participation also prohibits participation of any person who represents claimants and insurers. This aspect of the rule sometimes becomes the subject of controversy, particularly when one of the parties involved in the accident does not have the technical resources to commit to a major investigation, but at the same time wishes to participate. Such parties may procure specialists from outside sources, such as consulting firms with specialized expertise in a certain discipline. The controversy arises from the fact that quite often such

---

[4] Ref. Title 49 Transportation—Chapter VIII, Part 831 Paragraph 831.9 "Parties to the Field Investigation."

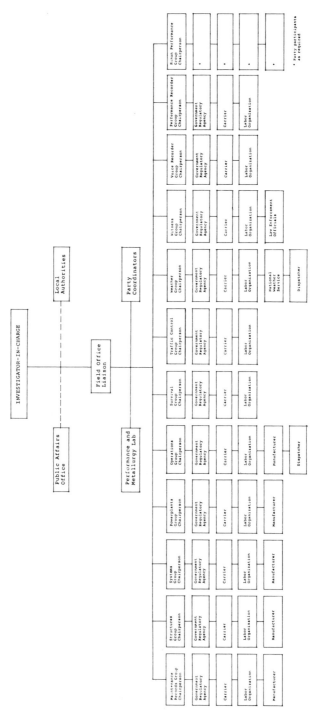

Figure 7.2 Major accident investigation organization, showing typical "team."

consultants may also be available to insurers and claimants. The NTSB makes every effort to keep accident investigations under its direction from becoming intertwined with other investigations involving legal or fault-finding aspects of the incident.

Separation of the technical and safety functions from the fault-finding and litigation aspects is further enforced in the Transportation Safety Act of 1974 which states, in part, "No part of any report of the Board, relating to any accident or investigation thereof, shall be admitted as evidence or used in any action for damages growing out of any matter mentioned in such report or reports."[5]

With the application of the foregoing safeguards against misuse of the party system to gain legal advantages, the technical expertise of each of the investigating groups is augmented by highly qualified technical and engineering personnel from

The governmental regulatory agency

The company that operated the vehicle involved in the accident (i.e., airline, bus or railroad company, steamship company)

Manufacturers of the vehicle or vessel and its major components or special components involved in the accident (i.e., tires, navigation equipment or power plants)

Members of labor organizations, such as those representing mechanics, ground support, dispatching, or emergency services personnel

Operators of airports or other transportation facilities that may have had a bearing on the accident

Personnel from special training facilities who may have trained pilots, drivers, or operators of other transportation systems

Other entities which, in the judgment of the investigator-in-charge, have specialized expertise and knowledge to contribute to the investigation

7.3.2.b Members of the Typical Investigation Team

Ideally, a team member provided to the investigative group by the member's employer (a "party" to the investigation) should have some general and special qualifications, which include

Ability to function under stress in an emotionally charged atmosphere

Good physical and mental health

---

[5] Section 304(c) Transportation Safety Act of 1974 "Use of Reports as Evidence."

Ability to remain open minded and objective while providing
meaningful information to the group, even if it "makes the
company look bad"

Ability to work well with other highly skilled persons with
possible adverse interests

Ability to be assertive without dominating the group

High level of skill in his or her discipline, with extensive
experience and academic credentials

Ability to speak effectively in a group of persons with adverse
interests, presenting ideas clearly in simple language, without
being argumentative

Ability to write well and to interpret complex technical notes, to
detect errors and incomplete documentation of facts

Ability to listen, understand, and convey to others contents of
complex technical discussions

Ability to communicate effectively with all levels of the member's own organization, from the lowest level technician to the
chief executive officer

Members of a typical aviation accident investigation team, and
their qualifications and responsibilities, are listed here. This list can
readily be adapted, with minor changes, to transportation accidents
involving other transportation modes.

Company or Party Coordinator

Qualifications: A high-level executive manager with technical
and administrative experience and with authority to make decisions
regarding technical, fiscal, and administrative matters for his or her
organization.

Responsibilities: Acts as the senior company representative at the
accident site and directs the activities of all company personnel; serves as
the company coordinator and is responsible for assigning appropriately
qualified company personnel to as many of the NTSB investigating groups
as deemed necessary; communicates any findings relative to maintenance
or operational safety that are disclosed at the accident site to company
headquarters and makes recommendations for corrective actions.

Operations Group Member

Qualifications: Must be knowledgeable regarding all flight operational aspects of the make and model of aircraft involved in the accident;
should be a highly qualified line pilot or supervisory or check airperson;
must be knowledgeable in cockpit procedures, flight crew training and
proficiency checks, maintenance of flight crew records, dispatching and
weather dissemination procedures, medical certification requirements,
in-flight emergency procedures, and federal aviation regulations.

Responsibilities: Acts as a full-time participant in all activities of
the NTSB operations group in the investigation of the above aspects of the
accident.

Air Traffic Control Group Member

Qualifications: A line pilot or dispatcher with special training or qualifications regarding all air traffic control (ATC) procedures and regulations.

Responsibilities: Participates in interviews of ATC personnel; assures that all ATC personnel with any knowledge of the circumstances of the accident are interviewed, that the scope of the interview is adequate and appropriate records of such interviews are made; assures that all pertinent ATC voice recordings and records are obtained and included in the record of the investigation.

Weather Group Member

Qualifications: Special training in operational aspects of meteorology, aviation weather observation, and disseminating requirements.

Responsibilities: Participates in all interviews of the U.S. Weather Service, Contract U.S. Weather Service, FAA, dispatch or private communications personnel involved in weather observations or recording or disseminating such information; assures that the scope of such interviews is adequate and an appropriate record of the interviews is made; assures that all pertinent records, photographs, reports, and statements are obtained and included in the investigation record.

Survival Factor Group Member

Qualifications: Familiarity with as many of the following disciplines as possible: aircraft emergency evacuation and fire fighting procedures, flight crew emergency procedure training, crash/fire/rescue (CFR) activities, search and rescue, survivability aspects of aircraft interior/cockpit configuration, and crash injury dynamics.

Responsibilities: Participates in the interviews of all key CFR and law enforcement personnel involved in the emergency response and fire fighting and rescue efforts; assures that such interviews are adequate in scope and that an appropriate record is made; assures that photographs, tape recordings, and other relevant documents are included in the investigation record.

Power Plant Group Member

Qualifications: Heavy line maintenance and overhaul experience on large turbine and/or reciprocating power plants and associated systems; a valid FAA power plant mechanic's certificate, or engineering accreditation with specialized experience in propulsion engineering and failure analysis, along with acquired practical experience in line maintenance or overhaul of power plants.

Responsibilities: Participates in on-site and off-site examination and/or disassembly of power plants, power plant components, fuel and oil system components, related instrumentation, transmitters and transducers, analysis of fuel and oil samples, analysis of fuel additives, power plant tests and performance records and appropriate documentation of such work; assures the review of all maintenance records pertaining to the power plants; assures the proper conduct of metallurgical analysis of any power plant part exhibiting non-impact-related fractures.

Systems Group Member

Qualifications: A valid FAA A&P Mechanic's certificate and extensive experience in the maintenance, overhaul, or repair and functional testing of aircraft systems; or an aeronautical engineer or flight engineer with extensive practical experience in the design, maintenance, or testing of aircraft systems.

Responsibilities: Participates in on-site and off-site examination, disassembly, and testing of aircraft pneumatic, hydraulic, electrical, avionic, deicing, anti-icing, fire protection, oxygen, flight control, and flight guidance systems; assures appropriate documentation of such tests and examinations.

Structures Group Member

Qualifications: An aircraft mechanic with significant experience in the examination, inspection, and repair of primary structures on large transport aircraft; or an aeronautical engineer with extensive experience in the design, testing, inspection, and repair of primary structures on large transport aircraft.

Responsibilities: Participates in on-site and off-site examination and/or testing of aircraft structural components, primarily to determine if a preexisting or preimpact structural failure occurred; directs and participates in metallurgical examinations of any suspect or failed aircraft structural part, and assures proper documentation of such examinations and tests.

Records Group Member

Qualifications: Extensive experience in the preparation and maintenance of company aircraft maintenance records, associated computer systems, maintenance and inspection requirements, design of automated record-keeping systems, and understanding of FAA requirements relative to record-keeping systems.

Responsibilities: Participates in all phases of investigative examination of company records relative to the maintenance and airworthiness of the airplane involved in the accident, such as records of the manufacturer, overhaul facility, repair agency, or other facility involved in the repair or inspection of parts and components installed on the involved aircraft; participates in the examination of FAA records; such as service difficulty reports, proposed airworthiness directives, and general notices (GNOTS). *Note:* The investigative authority of the NTSB provides for examination and copying of all records. It does *not* provide for taking possession of records. The records group must ensure that no originals of company records are retained by investigators. Original records must remain in the custody of the company. These could play a vital part in future litigation arising as a result of the accident.[6]

---

[6] Ref. Title 49 Transportation—Chapter VIII Part 831 Paragraph 831.8.

Witness Group Member

Qualifications: Experience or training in the art of interviewing witnesses, either in law enforcement work, or psychology, or legal training, and knowledge of flight operations and aeronautical terminology.

Responsibilities: Participates in the interviewing of eye witnesses to the accident and ensures that adequate statements are obtained from the witnesses for inclusion in the investigation record; ensures that such interviews are conducted properly and that witnesses are not led by any member of the group in an effort to elicit statements either favorable to his or her organization, or unfavorable to any other party.

Human Performance Group Member

Qualifications: Familiarity with the operational aspects of flight, an intimate knowledge of a flight crew member's duties and practical execution of cockpit responsibilities, understanding of cockpit resource management and training in human psychology.

Responsibilities: In cases where a formal human performance group is activated, the member participates in the coordination, selection, and interviewing of company personnel, friends, and relatives of deceased or injured crew members, or the crew members, if they survived the accident. Human performance investigation is a relatively new science used in accident investigation in efforts to determine why people do the things they do, to understand the human–machine relationship and, most importantly, to determine what needs to be done to prevent a recurrence of an accident where human performance was a cause or contributing factor. While the company may not have a qualified participant for this group, the company may elect to engage outside expertise for this purpose. Because of the relative new state-of-the-art of this aspect of an accident investigation, the potential sensitivity of the issues, and the highly specialized field, it is quite important for all parties to be fully aware regarding the status of the human performance investigation, and to participate to the maximum extent possible.

Cockpit Voice Recorder Group Member (aviation accidents only)

Qualifications: A pilot, fully qualified and current on the equipment involved in the accident; sufficient familiarity with members of the flight crew to be able to identify their voices. (If surviving flight crew members are available for review of the cockpit voice recorder (CVR) tapes, the second requirement is not necessary.)

Responsibilities: This group may not be involved in the accident site investigation, but instead is dispatched to the NTSB Washington CVR Laboratory as soon as the CVR is recovered. The member participates in the CVR group's activities until a complete and accurate transcript is produced. Critical information obtained from the CVR must be immediately communicated to the company coordinator at the accident site for the guidance of all investigators.

Flight Data Recorder Group Member (other performance-monitoring devices)

Qualifications: An orientation toward flight operations and aircraft performance; an aeronautical engineer with experience in aircraft performance, or a line pilot with an engineering background.

Responsibilities: This group generally does not work at the accident site, but instead proceeds to the NTSB Washington Flight Data Recorder Laboratory as soon as the flight data recorder (FDR) is recovered. The group member monitors the NTSB technician's FDR readout, ensures its accuracy, and provides essential FDR parameters information to his or her organization's coordinator at the accident site, as quickly as possible, for the guidance of all investigative personnel. Continuous coordination with the performance group is imperative.

Performance Group Member

Qualifications: An aeronautical engineer with extensive practical experience in calculating and analyzing aircraft performance; or a company flight engineer or line pilot with performance or engineering training or experience.

Responsibilities: The person assigned to this group will not proceed to the accident site, but will stand by at his or her regular duty station for notice from the company coordinator if an aircraft performance group is activated. This determination may not be made for several days after the accident, or when it becomes apparent that aircraft performance may have been a factor. The member assists the group in completing a performance analysis and in finding answers to all questions regarding aircraft performance. Continuous coordination with the FDR and CVR groups is imperative.

Other Groups

Other groups may be formed at the discretion of the investigator-in-charge, and corresponding group members with appropriate expertise may be required. For example, in the Air Florida accident reviewed in Section 7.4 it was determined that an in-depth investigation of airport operations and facilities was required; consequently an "airport group" was formed. In another case, early investigation disclosed the involvement of hazardous materials, and a "hazardous materials group" was created.

7.3.3 Procedures for the Investigative Team

After preliminary organization of the team, the investigator-in-charge will usually convene daily "progress meetings." These meetings are closed to the public and are attended only by those persons permitted to fully participate in the investigative process. An open review of all current factual findings and proposed direction of additional areas of inquiry are discussed.

The investigator-in-charge will also, in most cases, meet on a daily basis with party coordinators to review and resolve major issues, such as specific logistics or other subjects not appropriate for discussion with the entire investigation team.

It is the responsibility of all group members and coordinators to attend all NTSB progress meetings, ensure during such meetings that briefings by NTSB group chairpersons are accurate, and to speak up, without argumentation, to make appropriate corrections. The final and most important product of the individual group's investigative work is the "group notes." These notes will be the basis for the NTSB group chairperson's factual report, which will eventually be part of the official and public record of the investigation.

It is extremely important that each member of each group very carefully review a copy of the group notes before they are finalized by the NTSB group chairperson. This review must ensure

Absolute technical accuracy

Adequate description of the scope of the group's investigation

The absence of any irrelevant material involving aspects that did not contribute to the cause of the accident

Legibility of the notes

In no circumstances are group members to sign or initial group notes unless all of these four items have been accomplished. If a controversy regarding the notes cannot be resolved by the company group member and the NTSB group chairperson, the group member will immediately communicate this problem to the company coordinator, who in turn will affect a resolution of the problem with the NTSB investigator-in-charge.

When group notes are properly completed and reviewed, the NTSB group chairperson will provide copies of such notes to all members of the group and to each party or company coordinator. In no circumstances should the group's activity terminate or the group disband before all parties have received copies of group notes.

During the entire course of the investigation, on-site or off-site, it is of the utmost importance that all company personnel, particularly those performing as group members, keep their company coordinator completely informed regarding the status of the group's findings. This can best be accomplished at the company's debriefing sessions.

7.3.4 The Public Hearing: Applying the Party Concept

7.3.4.a *Public Hearing: A Continuing Phase of the Investigation (Figure 7.3)*

In most cases involving a catastrophic accident, regardless of the transportation mode, the NTSB requires a public hearing. The stated

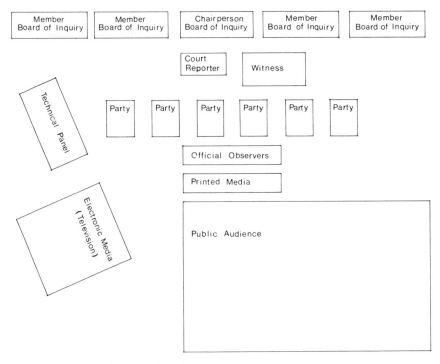

Figure 7.3 Organization of typical public hearing.

purpose of such hearings is described in the NTSB "Rules of Practice in Transportation Accident/Incident Hearings and Reports."[7]:

> Transportation accident hearings are convened to assist the Board in determining cause or probable cause of an accident, in reporting the facts, conditions and circumstances of the accident, and in ascertaining measures which will tend to prevent accidents and promote transportation safety. Such hearings are fact-finding proceedings with no formal issues and no adverse parties and are not subject to the provisions of the Administrative Procedure Act.

The philosophy of party participation in the public hearing phase of the investigation is essentially the same as during the "field" phase. Policy does not permit direct participation as a spokesperson by persons occupying legal positions or those representing claimants and insurers.

> The Chairman of the Board of Inquiry shall designate as parties to the hearing those persons, agencies, companies and associations whose participation in the hearing is deemed necessary in the public interest and whose special knowledge will contribute to

---

[7] Ref. Title 49 Transportation—Chapter VIII Part 845 Paragraph 845.2 "Nature of Hearing" Rev. 3/3/86.

the development of pertinent evidence. Parties shall be represented by suitable qualified technical employees or members who
do not occupy legal positions. . . . No party shall be represented
by any person who also represents claimants or insurers. Failure
to comply with this provision shall result in loss of status as a
party.[8]

This rule is a matter of occasional controversy, particularly in
cases of small operators/manufacturers or parties from foreign countries
who are not fully conversant with U.S. procedures or who are not fluent in
the English language. In such cases, parties may request the right to retain a
person other than an employee of their organization to act as a spokesperson at the public hearing; such requests are often denied by the NTSB, but
can be subject to judicial review.

7.3.4.b *Designation of Parties to the Hearing*

The process of designating parties to the hearing is somewhat
more formal than the designation of parties to the investigation, and is
usually done in writing by the chairperson of the board of inquiry.[9] Parties
to the hearing may, in many cases, be the same organizations that were
parties to the investigation. However, the fact that an organization was or
was not a party to the investigation does not automatically qualify or
disqualify such an organization for party status at the public hearing.

7.3.4.c *Introduction of Evidence into the Record:*
Witness Questioning Procedures

All of the factual material, such as group chairpersons' factual
reports, witness statements, and any other documentary evidence, is
introduced into the record at the public hearing and, in many cases, forms
the basis for eliciting additional answers from witnesses.

The NTSB rules regarding examination of a witness state that[10]

Any person who appears to testify at a public hearing shall be
accorded the right to be accompanied, represented or advised by
counsel or by any other duly qualified representative.

Witnesses shall be initially examined by the board of inquiry or its
technical panel. Following such examination, parties to the
hearing shall be given the opportunity to examine such witnesses.

Materiality, relevancy and competency of witness testimony,
exhibits or physical evidence shall not be the subject of objections
in the legal sense by a party to the hearing or any other person.
Such matters shall be controlled by rulings of the chairman of the
board of inquiry on his own motion. If the examination of a

---

[8] Ref. Title 49 Transportation—Chapter VIII Part 845 Paragraph 845.13.
[9] Title 49 Transportation—Chapter VIII Part 845 Paragraph 845.11 Rev. 3/3/86.
[10] Ref. Title 49 Transportation—Chapter VIII Part 845 Paragraphs 845.24 and 845.25
Rev. 3/3/86.

witness by a party is interrupted by a ruling of the chairman of the
board of inquiry, opportunity shall be given to show materiality,
relevancy or competency of the testimony or evidence sought to
be elicited from the witness.

The number of witnesses at a public hearing investigating a
catastrophic transportation accident may vary from 10 to more than 30,
depending on the number and complexity of the issues. The duration of the
hearing can also vary accordingly, from 2 or 3 days to 3 weeks.

The witnesses are first questioned by members of the technical
panel. The technical panel is usually headed by the investigator-in-charge
who was responsible for the field phase of the investigation. The investigator-in-charge is assisted by several of the key NTSB specialists (group
chairpersons) who headed the team's investigating groups responsible for
the development of evidence pertaining to the most critical areas of the
investigation.

After questioning by the technical panel members is completed,
the chairperson of the board of inquiry calls on each party to the hearing
and provides them with an opportunity to question the witness. The
chairperson does not permit the party spokesperson to ask repetitive,
needlessly complex questions or to cross-examine the witness. There are,
at times, tendencies for party spokespersons to rephrase questions that
have previously been answered, but in a manner not favorable to the
party's position. Repeated attempts by party spokespersons to ask such
questions over and over, phrased differently, in an effort to obtain answers
that may be more favorable, should be quickly terminated by the chairperson of the board of inquiry. The chairperson's rulings regarding the
admissibility of evidence and regulation of the course of the hearing are
final. Final questioning of the witness is done by the board of inquiry and
its chairperson. Additional rounds of questioning may be allowed by the
chairperson if new issues arise.

7.3.4.d   Compilation and Use of the Evidence

A court reporter makes a verbatim transcript of the entire public
hearing. This transcript, along with all other factual documentation, is
placed in the public docket and evaluated by the Board's staff.

7.3.5   The Report

After evaluation of all factual material, including the public
hearing transcript, each group chairperson prepares an analytical report.
These reports provide the writer of the final investigation report with the
individual group chairpersons' opinions as to the meaning of the facts as
they relate to the accident cause and to the prevention of similar accidents
in the future. Recommendations to prevent recurrences or involving
matters of safety are usually made as soon as the facts are determined. For

example, in the NTSB investigation of the Air Florida Potomac River accident of January 1982 (Section 7.4) ten major recommendations that could prevent a similar accident were issued to the Federal Aviation Administration within 15 days of the accident. Eleven additional recommendations were eventually issued to the FAA, along with the final report, on August 10–11, 1982.[11]

A discussion of the case history of the Air Florida accident will illustrate application of the team/group/party concept in an investigation.

7.4 A CASE HISTORY: AIR FLORIDA POTOMAC RIVER ACCIDENT

7.4.1 Description of the Accident

On January 13, 1982, Air Florida Flight 90, a Boeing 737-222 was scheduled for a flight to Fort Lauderdale, Florida, from Washington National Airport, Washington, DC. There were 74 passengers, including 3 infants and 5 crew members, on board. The flight's scheduled departure time was delayed about 1 hour 45 minutes because of a moderate to heavy snowfall, which necessitated temporary closure of the airport.

After takeoff from runway 36, which was made with reduced takeoff thrust and with snow and/or ice adhering to the aircraft, the aircraft crashed at 1601 (Eastern Standard Time) into the barrier wall of the northbound span of the 14th Street Bridge, which connects the District of Columbia with Arlington County, Virginia, and plunged into the ice-covered Potomac River. The aircraft came to rest on the west side of the bridge, 0.75 nautical mile from the departure end of runway 36. Four passengers and one crew member survived the crash. When the aircraft hit the bridge, it struck seven occupied vehicles and then tore away a section of the bridge wall and bridge railing. Four persons in the vehicles were killed; four were injured.

7.4.2 The Investigation

A complete discussion of the investigation of this case is beyond the scope of this chapter. Because of the multitude of issues involved and the voluminous documentation and analyses required, only selected portions of the individual group work is discussed here. This condensed discussion provides the reader with a brief overview of the scope of the investigative work accomplished by each discipline.[12]

---

[11] NTSB—AAR-82-8 (Government Accession No. PB82-910408) "Air Florida Inc., Boeing 737-222, N62AF, Collision with 14th Street Bridge, near Washington National Airport, Washington, DC, January 13, 1982."

[12] Certain parts of Section 7.4 are extracted verbatim from the NTSB formal report (footnote no. 11).

7.4.2.a *Operations Group Investigation*

The operations group assembled factual data on the sequence of all operations including preflight preparation, passenger loading, deicing, departure from the gate, taxi and pre-takeoff, takeoff, and impact. Excerpts from the group report include the following:

Air Florida, Inc. Flight 90, a Boeing 737-222, was a scheduled passenger flight from Washington National Airport, Washington, DC, to Fort Lauderdale International Airport, Fort Lauderdale, Florida, with an intermediate stop at Tampa National Airport, Tampa, Florida. Flight 90 was scheduled to depart Washington National Airport at 1415 e.s.t.[13] The aircraft had arrived at gate 12 as Flight 95 from Miami at 1329. Snow was falling in Washington, DC, in the morning and in various intensities when flight 95 landed and it continued to fall throughout the early afternoon. Because of the snowfall, Washington National Airport was closed for snow removal from 1338 to 1452, necessitating the delay of Flight 90's scheduled departure. At 1359, Flight 90 requested and received an instrument flight rules (IFR) clearance from clearance delivery.

Seventy-one passengers and three infants were boarded on the aircraft between 1400 and 1430; there were five crew members—the captain, the first officer, and three flight attendants. About 1420, American Airlines[14] maintenance personnel began deicing the left side of the fuselage. Equipment and solutions used for the deicing were noted. The deicing truck operator stated that the captain requested deicing just before the airport was scheduled to reopen at 1430 so that he could get in line for departure. American maintenance personnel stated that they observed about one-half inch of wet snow on the aircraft prior to application of the deicing fluid. Fluid had been applied to an area of about 10 square feet when the captain terminated the operation because the reopening of the airport had been delayed. At that time, the flight crew also informed the Air Florida maintenance representative that 11 other aircraft had departure priority and that there were 5 to 6 aircraft that had departure priority before Flight 90 could push back from the gate.

Between 1445 and 1450, the captain requested that the deicing operation be resumed. The left side of the aircraft was deiced first. The operator of the deicing vehicle stated that the wing, fuselage, tail section, top part of the engine pylon, and cowling were deiced with a heated solution consisting of 30–40% glycol and 60–70% water. No final over-spray was applied. The proportions of the mix were based on the American Airlines maintenance manual and on the operator's knowledge that the ambient air temperature was 24°F. He had obtained this temperature from current weather data received at the American Airlines line maintenance room. The operator also stated that he started spraying at the front section

---

[13] All times used are Eastern Standard Time based on the 24-hour clock.
[14] American Airlines had contracted to provide certain services to Air Florida.

of the aircraft and progressed toward the tail, using caution in the areas of the hinge points and control surfaces to ensure that no ice or snow remained at these critical points. He stated that it was snowing heavily as the deicing/anti-icing substance was applied to the left side of the aircraft. Between 1445 and 1500, the operator of the deicing vehicle was relieved from his task. He told his relief operator, a mechanic, that the left side of the aircraft had been deiced.

The relief operator proceeded to de-ice the right side of the aircraft with heated water, followed by a heated finish anti-ice coat of 20–30% glycol and 70–80% water. These proportions were based on information that the ambient temperature was 28°F. (The actual temperature was 24°F.) The operator stated that he deiced in the following sequence: rudder, stabilizer and elevator, aft fuselage section, upper forward fuselage, wing section (leading edge to trailing edge), top of the engine, wingtip, and nose. Afterward, he inspected both engine intakes and the landing gear for snow and/or ice accumulation; none was found. The deicing/anti-icing was completed at 1510. At this time about 2 to 3 inches of wet snow lay on the ground around the aircraft. Maintenance personnel involved in the deicing operation stated they believed that the aircraft's trailing and leading edge devices were retracted. American Airlines personnel stated that no covers or plugs were installed over the engines or airframe openings during deicing operations.

At 1515, the aircraft was closed and the jetway was retracted. Just prior to retraction of the jetway, the captain, sitting in the left cockpit seat, asked the Air Florida station manager, standing near the main cabin door, how much snow was on the aircraft. The station manager responded that there was a light dusting of snow on the left wing from the engine to the wingtip and that the area from the engine to the fuselage was clean. Snow continued to fall heavily.

A tug was standing by to push Flight 90 back from gate 12. At 1516, Flight 90 requested clearance from the tower to push back and get in sequence for takeoff. At 1523, clearance was given for the pushback operation. The tug attempted to push the aircraft back at 1525, but it was not equipped with tire chains and a combination of ice, snow, and glycol on the slightly inclined ramp prevented the tug from moving the aircraft. A flight crew member suggested to the tug operator that the aircraft engines' reverse thrust could be used for backing the aircraft from the gate. The tug operator advised the crew member that this suggestion was contrary to the policy of American Airlines. According to the tug operator, the aircraft's engines were started and both reversers were deployed. He then advised the flight crew to use only "idle power."

Witnesses estimated that both engines were operated in reverse thrust for a period of 30 to 90 seconds. During this time several Air Florida and American Airlines personnel observed snow and/or slush blown toward the front of the aircraft. One witness stated that he saw water swirling at the base of the left (No. 1) engine inlet. Several Air Florida personnel stated that they saw an area of melted snow on the ground around the left engine for a radius ranging from 6 to 15 feet. No one observed a similar melted area under the right (No. 2) engine.

When the use of reverse thrust proved unsuccessful in moving the aircraft back, the engines were shut down with the reversers deployed. The same American Airlines mechanic that had inspected both engine intakes earlier performed another general examination of both engines. He stated that he saw no ice or snow at that time. Maintenance personnel standing near the aircraft after the engines were shut down stated that they did not see any water, slush, snow, or ice on the wings.

At 1533, while the first tug was being disconnected from the towbar and a second tug was brought into position, an assistant manager for Air Florida stated that he could see the upper fuselage and about 75% of the left wing from his vantage point, which was inside the terminal about 25 feet from the aircraft. Although he observed snow on top of the fuselage, he said it did not appear to be heavy or thick. He saw snow on the nose and radome up to the bottom of the windshield and a light dusting of snow on the left wing. At 1535, Flight 90 was pushed back without further difficulty. After the tug was disconnected, both engines were restarted and the thrust reversers were stowed. The aircraft was ready to taxi away from the gate at 1538.

At 1538:16, while accomplishing "after-start" checklist items, the captain responded "off" to the first officer's callout of checklist item "anti-ice." At 1538:22, Flight 90 responded to ground control inquiries by stating the aircraft was ready to taxi. Ground control transmitted "Okay Palm Ninety, Roger, just pull up over behind that . . . TWA and hold right there.[15] You'll be falling in line behind a . . . Apple . . . DC Nine."[16] Flight 90 acknowledged this transmission and fell in line behind the New York Air DC-9. Nine air carrier aircraft and seven general aviation aircraft were awaiting departure when Flight 90 pushed back.

At 1540:15, the cockpit voice recorder recorded a comment by the captain, "Go over to the hanger and get deiced," to which the first officer replied, "Yeah, definitely." The captain then made some additional comment that was not clear but contained the word *deiced*, to which the first officer again replied "Yeah—that's about it." At 1540:42 the first officer continued, "It's been awhile since we've been deiced." At 1546:21 the captain said, "Tell you what, my windshield will be deiced, don't know about my wings." The first officer then commented, "Well—all we need is the inside of the wings anyway, the wingtips are gonna speed up on eighty anyway, they'll shuck all that other stuff." At 1547:32 the captain commented, "Gonna get your wing now." Five seconds later the first officer asked, "D'they get yours? Did they get your wingtip over'er?" The captain

---

[15] *Palm 90*—ATC designation for Air Florida Flight 90.
[16] *Apple*—ATC designation for New York Air Flights.

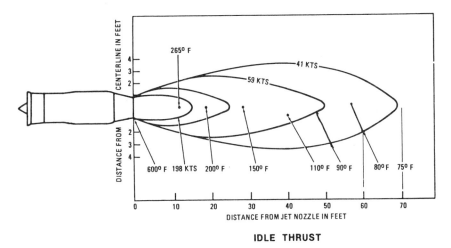

IDLE THRUST

Figure 7.4 Jet blast profile for Pratt & Whitney JT8-D engine (KTS = knots).

replied, "I got a little on mine." The first officer then said, "A little, this one's got about a quarter to half an inch on it all the way."

At 1548:59 the first officer asked, "See this difference in that left engine and right one?" The captain replied "Yeah." The first officer then commented, "I don't know why that's different—less it's hot air going into that right one, that must be it—from his exhaust—it was doing that at the chocks[17] awhile ago." At 1551:54, the captain said, "Don't do that—Apple, I need to get the other wing done" (Figure 7.4).

At 1553:21 the first officer said, "Boy . . . this is a losing battle here on trying to deice those things, it (gives) you a false feeling of security that's all that does." Conversation between the captain and first officer regarding the general topic of deicing continued until 1554:04.

At 1557:42 after the New York Air aircraft was cleared for takeoff, the captain and first officer proceeded to accomplish the pre-takeoff checklist, including verification of the takeoff engine pressure ratio (EPR)[18] setting of 2.04 and indicated airspeed bug[19] settings of 138 knots (V_1),[20] 140

---

[17] Chocks—blocks placed in front and/or behind the wheels to prevent a parked aircraft from moving; also the designation for the area where the aircraft is parked for passenger and cargo loading.

[18] Engine pressure Ratio (EPR)—a reading on flight deck instruments of engine thrust derived by measuring the ratio of the pressure at the engine's turbine discharge point versus the compressor inlet pressure.

[19] Airspeed bug—A movable pointer on the flight deck instrument panel air speed indicator that is set at certain calculated and predicted airspeeds for different flight regimes.

[20] Takeoff decision speed—The speed at which, if an engine failure occurs, the distance to continue the takeoff to a height of 35 feet will not exceed the usable takeoff distance, or the distance to bring the airplane to a full stop will not exceed the acceleration–stop distance available.

knots (V_R),[21] and 144 knots (V_2).[22] Between 1558:26 and 1558:37 the first officer, who was flying the aircraft, asked, "Slush runway, do you want me to do anything special for this or just go for it?" The captain responded, "Unless you got anything special you'd like to do." The first officer replied, "Unless just take off the nosewheel early like a soft field takeoff or something; I'll take the nosewheel off and then we'll let it fly off."

At 1558:55 Flight 90 was cleared by local control to "taxi into position and hold" on runway 36 and to "be ready for immediate (takeoff)." As the aircraft was taxied, the tower transmitted takeoff clearance and the pilot acknowledged, "Palm 90 cleared for takeoff." Also, at 1559:28 Flight 90 was told not to delay the departure, as landing traffic was $2\frac{1}{2}$ miles out for runway 36. The last radio transmission from Flight 90 was the reply "okay" at 1559:46.

The CVR indicated that the pre-takeoff checklist was completed at 1559:22. At 1559:45, as the aircraft was turning to the runway heading, the captain said "Your throttles." At 1559:46 the sound of engine spoolup was recorded and the captain stated, "Holler if you need the wipers." At 1559:56 the captain commented, "real cold, real cold," and at 1559:58 the first officer remarked, "God, look at that thing, that don't seem right, does it?"

Between 1600:05 and 1600:10 the first officer stated, "that's not right," to which the captain responded, "Yes it is, there's eighty." The first officer reiterated, "Naw, I don't think that's right." About nine seconds later the first officer added "maybe it is," but then two seconds later, after the captain called "hundred and twenty," the first officer continued, "I don't know."

Eight seconds after the captain called "Vee one" and two seconds after he called "Vee two," the sound of the stickshaker[23] was recorded. At 1600:45 the captain said "Forward, forward," and at 1600:48 "We only want five hundred." At 1600:50 the captain continued, "Come on, forward, forward, just barely climb." At 1601:00 the first officer said, "Larry, we're going down, Larry," to which the captain responded, "I know it."

About 1601, the aircraft struck the heavily congested northbound span of the 14th Street Bridge, which connects the District of Columbia with Arlington County, Virginia, and plunged into the ice-covered Potomac River. It came to rest on the west end of the bridge 0.75 nautical mile from the departure end of runway 36. Heavy snow continued to fall and visibility at the airport was varying between $\frac{1}{4}$ and $\frac{5}{8}$ mile.

When the aircraft struck the bridge, it struck six occupied

---

[21] *Rotation speed*—the speed at which rotation is initiated during the takeoff to attain climb speed at the 35-foot height.

[22] *Climb speed*—The scheduled target speed to be attained at the 35-foot height.

[23] Stickshaker—A device that activates to warn the flight crew of an impending stall by artificially setting up vibration in the control column accompanied by an aural warning.

automobiles and a boom truck before tearing away a 41-foot section of the bridge wall and 97 feet of the bridge railings. As a result of the crash, 70 passengers, including 3 infants, and 4 crew members were killed. Four passengers and one crew member were injured seriously. Four persons in vehicles on the bridge were killed; four were injured, one seriously.

7.4.2.b Witness Group Report

Investigators interviewed more than 200 witnesses to establish the sequence of events from the start of takeoff until impact. More than 100 written statements were obtained. (See Figure 7.5 for witness locations and flight path. Numbers 1 through 10 indicate locations of the witnesses interviewed.)

Ground witnesses generally agreed that the aircraft was flying at an unusually low altitude, with the wings level, and that a nose-high attitude of 30° to 40° was obtained before impact with the bridge (Figure 7.6). Four persons in a car on the bridge within several hundred feet from the point of impact claimed that large sheets of ice fell on their car.

The following description was provided by a driver whose car on the bridge was at about the wingtip of the aircraft:

> I heard screaming jet engines . . . the nose was up and the tail was down. It was like the pilot was still trying to climb but the plane was sinking fast. I was in the center left lane . . . about 5 or 6 car lengths from where (the red car) was. I saw the tail of the plane tear across the top of the cars, smashing some tops and ripping off others. . . . I saw it (the red car) spin around and then hit the guardrail. All the time (the plane) was going across the bridge it was sinking but the nose was pretty well up. . . . I got the impression that the plane was swinging around a little and going in a straight direction into the river. The plane . . . seemed to go across the bridge at a slight angle and the dragging tail seemed to straighten out. It leveled out a little. Once the tail was across the bridge the plane seemed to continue sinking very fast but I don't recall the nose pointing down. If it was, it wasn't pointing down much. The plane seemed to hit the water intact in a combination sinking/plowing action. I saw the cockpit go under the ice. I got the impression it was skimming under the ice and water. . . . I did not see the airplane break apart. It seemed to plow under the ice. I did not see any ice on the aircraft or any ice fall off the aircraft. I do not remember any wing dip as the plane came across the bridge. I saw nothing fall from the airplane as it crossed the bridge.

Between 1519 and 1524, a passenger on an arriving flight holding for gate space near Flight 90 saw some snow accumulated on the top and right side of the fuselage, and took a photograph of the aircraft.

No witnesses saw the flight crew leave the aircraft to inspect for snow/ice accumulations while at the gate. Departing and arriving flight crews and others who saw Flight 90 before and during takeoff stated that the aircraft had an unusually heavy accumulation of snow and ice on it. An airline crew taxiing parallel to Flight 90 during its takeoff, but in the

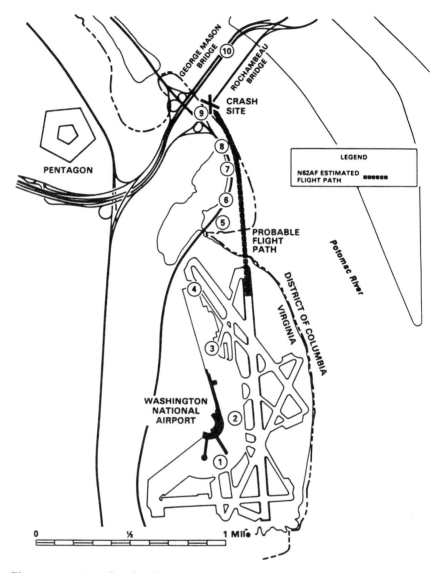

Figure 7.5 Air Florida Flight 90 probable flight path (reconstructed) and location of key witnesses.

opposite direction, saw a portion of Flight 90's takeoff roll and discussed the extensive amount of snow on the fuselage. The captain's statement to the board included the following: "I commented to my crew, 'Look at the junk on that airplane.' Almost the entire length of the fuselage had a mottled area of snow and what appeared to be ice . . . along the top and

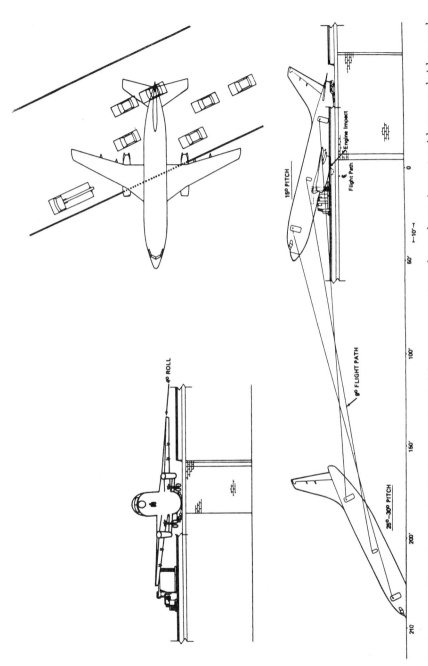

Figure 7.6 Aircraft impact attitude, reconstructed from examination of wreckage, impact evidence on bridge, and witness statements.

upper side of the fuselage above the passenger cabin windows." None of the witnesses at the airport could positively identify the rotation or liftoff point of Flight 90; however, they testified that it was beyond the intersection of runways 15 and 36, and that the aircraft's rate of climb was slow as it left the runway. Flight crews awaiting departure were able to observe only about the first 2000 feet of the aircraft's takeoff roll because of the heavy snowfall and restricted visibility.

One witness at the airport stated, "Immediately after I noticed the Air Florida 737, an Eastern 727 landed unbelievably close after the (Air Florida) 737. I felt it was too close for normal conditions—let alone very hard snow."

7.4.2.c Weather Group Report

The following terminal forecast was issued by the National Weather Service Forecast Office, Washington, DC, at 0940 on January 13, and was valid 1000, January 13 through 1000, January 14, 1982:

Ceiling—1500 feet overcast, visibility—3 miles reduced by light snow, variable ceiling 500 feet obscured, visibility—$\frac{3}{4}$ miles reduced by light snow. After 1300: ceiling 600 feet obscured, visibility—1 mile reduced by light snow, wind—130° 10 knots, occasionally ceiling 300 feet obscured, visibility—$\frac{1}{2}$ miles reduced by moderate snow. After 1700: ceiling 400 feet obscured, visibility—1 mile reduced by light snow, occasional visibility—$\frac{1}{2}$ miles reduced by moderate snow, chance of light freezing rain, light ice pellets and moderate snow. After 0100: ceiling 1500 feet overcast, visibility—4 miles reduced by light snow, wind—310° 10 knots. After 0400: marginal visual flight rules due to ceiling and snow.

The following SIGMET ALPHA-3[24] was issued at 1347 on January 13 by the National Weather Service, Washington, D.C., which was valid from 1340 through 1740 for Ohio, West Virginia, Virginia, District of Columbia, Maryland, and Delaware:

From 20 miles northwest of Erie to 60 miles northeast of Parkersburg to Atlantic City to Hatteras to Savanna to 60 miles east of Chattanooga to York (Kentucky) to Cincinnati. Moderate occasional severe rime or mixed icing in clouds and in precipitation above the freezing level reported by aircraft. Freezing level from the surface over Ohio sloping to multiple freezing levels from the surface to 6000 feet over central Carolinas, southeast Virginia and the Delmarva Peninsula. Freezing level 7000 to 9000 feet over the coastal Carolinas. Continue advisory beyond 1740.

---

[24] *Significant meteorological information (SIGMET)*—A weather advisory issued concerning weather significant to the safety of all aircraft. SIGMET advisories cover severe and extreme turbulence, severe icing, and widespread dust or sandstorms that reduce visibility to less than 3 miles.

The following surface observations were taken before and after the accident by observers under contract to the National Weather Service (NWS) at Washington National Airport:

> 1558: type—record special; ceiling—indefinite 200 feet obscured; visibility—½ mile; weather—moderate snow; temperature—24° F; dewpoint—24° F; wind—010° 11 knots; altimeter—29.94 inches; remarks—runway 36 visual range 2800 feet, variable 3500 feet.
>
> 1614: type—special; ceiling—indefinite 200 feet obscured; visibility—¾ mile; weather—moderate snow; temperature—24° F; dewpoint—24° F; wind—020° 13 knots; altimeter—29.91 inches; remarks—runway 36 visual range 2000 feet, variable 3500 feet, pressure falling rapidly (aircraft mishap).

The precipitation intensities recorded before and after the accident were as follows:

	Time	
Precipitation	Began	Ended
Moderate snow	1240	1320
Heavy snow	1320	1525
Moderate snow	1525	1540
Light snow	1540	1553
Moderate snow	1553	1616

The following are the synoptic observations of precipitation water equivalent and measured snow accumulation:

Time (from–to)	Water Equivalent	Snow Accumulation
0650–1252	0.07 inches	2.1 inches
1252–1851	0.32 inches	3.8 inches
Midnight–midnight	0.42 inches	6.5 inches

There were two transmissometers in operation before and during the time of the accident. The center of the baseline of the transmissometer for runway 36 was located about 1600 feet down the runway from the threshold and about 600 feet to the right of the runway centerline. The center of the baseline for runway 18 was located about 1700 feet down the runway from the threshold and about 800 feet to the left of the runway centerline. Both transmissometers had a 250-foot baseline.

Runway visual range[25] was measured as follows for the times indicated:

| Time | Runway Visual Range (feet) | |
	Runway 36	Runway 18
1544	3800	2300
1558	2900	1500
1600	2100	1400
1604	1800	1200
1610	2900	1600

Air Florida Flight 90 received weather briefing information from American Airlines at Washington National Airport. The operations agent stated that copies were not kept of weather information or a log of what was delivered to the flight crew. In a written statement the operations agent noted that in addition to destination information, Air Florida Flight 90 would have received current surface observations at Washington National Airport (excluding a field condition report).

7.4.2.d *Air Traffic Control Group Report*

The ATC group investigated the following areas: ATC information to pilots, separation criteria, controller experience, and gate-hold procedures.

The FAA *Air Traffic Control Handbook* (7110.65B) requires that runway visual range (RVR) or runway visibility value (RVV)[26] be issued for runways in use when the prevailing visibility is 1 mile or less, regardless of the value indicated or "when RVR/RVV indicates a reportable value regardless of the prevailing visibility." Also required is the issuance of "mid-rollout RVR when the value of either is less than 2000 feet and less than the touchdown value." Neither RVR nor RVV was issued by air traffic control to Flight 90. The RVR, however, was issued to landing aircraft as they were cleared to land.

The Automatic Terminal Information Service (ATIS) is a continuous broadcast of recorded noncontrol information in selected terminal areas. It is intended to improve controller effectiveness and to relieve frequency congestion by automating the repetitive transmission of essential, but routine, information such as weather conditions, runway conditions, temperatures, and altimeter settings. Pilots are expected to monitor

---

[25] *Runway visual range (RVR)*—An instrumentally derived value that represents the horizontal distance a pilot will see down the runway from the approach end.

[26] Runway visibility value (RVV)—The visibility determined for a particular runway by a transmissometer.

ATIS preliminary to departure from or arrival at an airport and to advise ATC of the code of the ATIS message. The FAA *Facility Operation and Administration Manual* (7110.3F) requires that messages be brief, not to exceed 30 seconds unless required for message content completeness, and that each message be identified by a phonetic alphabet letter code word at both the beginning and end of the message. A new recording is to be made on receipt of any new official weather, whether there is or is not a change of values; a new recording is also to be made when there is a change in any other pertinent data, such as runway change, instrument approach in use, or new or canceled SIGMETS. On the day of the accident, ATIS information was not updated with changes in braking action.

ATIS "Alpha" was broadcast from 1514 to 1531. Braking action had been reported as "fair" by multiengine commuter aircraft. Ground control also received braking reports at 1511 as "poor, especially at turnoff" from a U.S. Air BAC-111 and "fair to poor" from an Eastern DC-9 aircraft. ATIS "Bravo" was broadcast from 1532 to 1537 and contained no braking report. Subsequent "Bravo" broadcasts from 1538 to 1544 and from 1544 to 1602 listed braking as "poor" as reported by a B-727 aircraft. According to the requirements of the FAA manual, the second and third "Bravo" broadcasts should have been coded "Charlie" and "Delta," respectively. Flight 90 did not inform clearance delivery or ground control personnel that it was in receipt of ATIS, nor were the flight crew asked by either clearance delivery or ground control about receipt of such information.

Criteria for separation between departing and arriving aircraft are set forth in the FAA *Air Traffic Control Handbook* (7110.65B) as follows:

> Except as provided in 744, separate a departing aircraft from an arriving aircraft on final approach by a minimum of 2 miles if separation will increase to a minimum of 3 miles (5 miles when 40 miles or more from the antenna) within 1 minute after takeoff.

The FAA Handbook requires that the controller determine the position of an aircraft before issuing taxi information or takeoff clearances. Such position may be determined visually by the controller, by pilot reports, or by the use of airport surface detection radar equipment. With regard to Flight 90, because of limited visibility, the local controller could not see the aircraft when he cleared the flight by issuing the messages "into position and hold" and "be ready for an immediate takeoff." There is no airport surface detection equipment at the Washington National Airport control tower.

Using the ATC tape, FDR readout from Eastern Flight 1451, radar data from Eastern 1451, and radar performance data for Flight 90, the Board calculated the distance between Flight 90 and the landing aircraft, Eastern 1451. The distance was calculated to be between 1500 and 4000 feet. Discrepancies in the FDR from Flight 1451 precluded more precise calculations.

The FAA *Air Traffic Control Handbook* (7110.65B) stresses in a Traffic Training Program Lesson Plan *not "to clear a departure for takeoff when the arrival is 2 miles from the runway; it is too late then. Normally, departure action must be taken at 3 miles to realize a 2 mile minimum."* Additionally, this provision is contained in the written Local Control Test No. 1, which was given to Washington National Airport controllers as part of their initial training.

Controller staffing levels and experience were reviewed with the following results. On January 13, 1982, the local control, ground control, clearance delivery, and departure control positions were occupied. This staffing represents a full complement, identical to that of July 1981, before the controllers' strike. Ground control, clearance delivery, and departure control positions were occupied by developmental controllers who had all been checked out in their respective positions.

The local controller handling Flight 90 at the time of the accident was a working controller and was also the team supervisor. He began his career as a military controller in 1959 and had worked at Washington National Airport since 1964. His training records indicated that he was checked out on all operating positions. His training file indicated that his last "over the shoulder" check was completed satisfactorily on September 9, 1977. The "over the shoulder" training review is required to be administered semiannually. Testimony at the Safety Board public hearing indicated that he had been given these required checks, but written documentation could not be provided.

The study of gate-hold procedures revealed that Washington National Tower did not use gate-hold procedures on the day of the accident. Gate-hold procedures were initially developed as a fuel conservation measure. The FAA *Air Traffic Control Handbook* (7210.3F) outlines these procedures as follows:

> The objective of gate hold procedures is to achieve departure delays of 5 minutes or less after engine start and taxi time. Facility chiefs shall ensure that gate hold procedures and departure delay information are made available to all pilots prior to engine start up. Implement gate hold procedures whenever departure delays exceed or are expected to exceed five minutes.
>
> Facility chiefs shall meet with airport managers and users to develop local gate hold procedures within the guidelines of 1230 and in accordance with limitations imposed by local conditions. Include the following general provisions in the procedures:
>
> 1. Pilots shall contact GC/CD (ground control/clearance delivery) prior to starting engines to receive start time. The sequence for departure shall be maintained in accordance with initial callup unless modified by flow control restrictions.
> 2. Develop notification procedures for aircraft unable to transmit without engine(s) running. Note: Inability to contact GC/CD prior to engine start shall not be justification to alter departure sequence.

3. The operator has the final authority to decide whether to absorb the delay at the gate, have the aircraft towed to another area, or taxi to a delay absorbing area.
4. GC/CD frequency is to be monitored by the pilot for a new proposed engine start time if the delay changes.

The chief of the Washington National Tower stated that, because of airport space limitations, gate-hold procedures could not be implemented and that Washington Tower Letter to Airmen 79-1 (subject: Departure Delay Procedures—Fuel Conservation) had been written to comply with the foregoing requirements. The letter went into effect November 20, 1979 and expired November 20, 1981. It had not been renewed.

Between 1517:13 and 1547:55, there were a total of 22 communications between ground control and aircraft on the ground at Washington National Airport relative to flight crew concerns over departure information. The tower was unable to provide these departing flight crews with reasonable estimates of anticipated departure delays.

7.4.2.e Airport Group Report

An airport group was formed some time after the initial investigating teams were organized. The investigator-in-charge had determined that this group was needed after it was indicated that airport design, location, maintenance, and facilities might be factors in the accident. This group studied the airport location, runways, and safety areas; airport emergency procedures (including snow removal); and rescue equipment. Excerpts from the report follow.

Washington National Airport is located at Gravelly Point, Virginia, on the west bank of the Potomac River. Arlington County, Virginia is to the immediate west, while the city of Alexandria, Virginia, is to the south. The east boundary of the airport is the Potomac River, while the District of Columbia is directly to the north. The areas surrounding the airport are populated, and the general center of Washington, DC, is about 3 miles north of the airport. Washington National Airport is owned by the U.S. government and is operated by the Federal Aviation Administration, U.S. Department of Transportation. Washington National Airport was opened in 1941.

The landing area consists of three runways: 18-36, 15-33, and 3-21. Runway 18-36 is served by a Category II instrument landing system (ILS), high-intensity runway lights, high-intensity approach lighting system with sequenced flashing lights, touchdown zone lights, and centerline lights. Runway 18-36 is hard surfaced with asphalt and grooved; it is 6869 feet long and 150 feet wide. Edge lights on runway 18-36 are displaced 35 feet on each side of the runway.

Runway 36 at Washington National Airport has a runway safety area (overrun) that complies with current FAA design criteria for existing runway safety areas. The design criteria require that the safety area be 500 feet wide and extend 200 feet beyond the end of the runway. The runway 36 safety area is 500 feet wide and 335 feet beyond the end of the runway.

However, FAA design criteria for newly constructed runways require an extended runway safety area in addition to the runway safety area. The extended runway safety area is that rectangular area along the extended runway centerline that begins 200 feet from the end of the usable runway (the 200-foot area is the runway safety area) and extends outward in conformance with criteria in effect at the time of construction. Current FAA criteria for new airports (constructed at the time of the investigation) require that the extended runway safety area be 800 feet long and 500 feet wide. The total length of the two safety areas must be 1000 feet beyond the end of the runway.

Emergency procedures and snow removal were studied. FAA Airport Bulletin DCA 7/45, dated October 9, 1981, contains snow and slush emergency procedures that were in effect from the date of issue through April 1, 1982. The purpose of the bulletin was to assign responsibilities and to establish procedures for removing and controlling snow, slush, ice, sand, and water at Washington National Airport.

At 1245 on January 13, airport personnel measured the snow on runway 18-36 and found it to be about 2 inches. Shortly thereafter, the airport operations office decided to remove the snow from the runway. At 1250, an airport advisory was issued: "Airport will be closed 1330–1430 for snow removal." Snow was to be removed using snow plows with rubber boots on the blades. Plows removed snow down to the surface. Brooms were used to sweep away any remaining loose snow after the plows passed, and the runway surface was then sanded.

At 1450, snow removal on runway 18-36 was completed and the airport was reopened. At this time, the air traffic control tower was notified, "runway 18-36 plowed full length and width, sanded 50 feet each side of centerline. All other surfaces covered with $3\frac{1}{4}$ inches of dry snow."

At 1525, the airport operations officer issued the following airport field report: "runway 18-36 plowed, swept full length and width, sanded 50 feet each side of centerline. All other surfaces covered with $3\frac{3}{4}$ inches of dry snow. Use caution." At 1600, an airport operations officer "estimated that the snow cover on the last 1500 feet of runway 36 amounted to about three-fourths inch." At 1600:22, the local controller made a general transmission, "brakes poor" on runway 36. At 1607, American Airlines Flight 508, a Boeing 727 aircraft, landed on runway 36, and the captain reported that braking action was "poor" and that snow was rapidly covering the runway.

Under requirements of 14 CFR 139, *Certification and Operations: Land Airports Serving CAB—Certified Air Carriers*, certified airports are required to provide primary crash/fire/rescue (CFR) protection within the geographical boundaries of the airport. There is no requirement to respond to accidents outside the boundaries of the airport.

Advisory Circular (AC) 150/5210-13, "Water Rescue Plans, Facilities and Equipment," dated May 4, 1972, suggests the planning procedures and necessary facilities and equipment to effectively perform rescue operations when an aircraft lands in a body of water, swamp, or tidal area, where normal aircraft fire fighting and rescue service vehicles are unable to reach the accident scene. The AC states that special water rescue services, where possible, should be under the jurisdiction of the airport management and located on or near the airport. In this and all other situations, the procedures should be coordinated with local emergency rescue services. With regard to vehicles, the AC states that air-cushion vehicles have high-speed capabilities over water and adverse terrain conditions, which make them ideally suited for rescue service. If this type of vehicle is available, its use should be included in the emergency rescue plan.

7.4.2.f *Power Plant Group Report*

The aircraft was equipped with two Pratt & Whitney JT8D-9A turbo-fan engines with a takeoff thrust rating of 14,500 pounds each at sea level on a standard day. Engine power settings for the Flight 90 takeoff from Washington National Airport were to be 2.04 EPR.

The engines were examined at the accident scene and their condition was documented. They were then taken to the facilities of Pratt & Whitney Aircraft Group, Hartford, Connecticut, for disassembly and inspection. No evidence of preimpact malfunction was noted. External and internal examination of both engines' high-pressure and low-pressure compressors and turbine sections disclosed varying degrees of damage consistent with rotation at impact. There was no evidence of any lack of lubrication on any bearings. The oil systems were not contaminated.

Each of the two engines was equipped with a thermal anti-ice system, composed of three anti-ice valves that were designed to open when the respective engine anti-ice switch is placed ON. The inlet guide vanes and nose cones use 8th stage compressor bleed air and the cowl anti-ice system uses 13th stage air. The 8th and 13th stage engine bleed air ducting on the right side of the left engine was crushed between the 1:30 and 4:00 positions. The left engine nose cowl thermal anti-ice valve was closed. The main bleed air valve was closed. The fuel heat valve was closed. The left inlet guide vane anti-ice valve was closed. The right inlet guide vane anti-ice valve was open and free to rotate. The air turbine starter was not visibly damaged.

The right engine right inlet guide vane anti-ice valve was closed. The left inlet guide vane anti-ice valve was not recovered. The nose cowl anti-ice valve was closed. The engine bleed valve was closed. The modulation/shutoff valve was closed. The fuel heat valve was closed. The air turbine starter exhibited no visible damage. The pressurization and bleed control was not visibly damaged. The control was disassembled and no mechanical discrepancies were noted, except that it was clogged with water and dirt. The 8th stage and two 13th stage antisurge bleed valves functioned normally.

Thrust-indicating system/engine instrument readings were recorded. Impact damage and loss of electrical power at impact are considered in making postimpact determination of instrument condition.

Both EPR transmitters had been damaged by impact. By comparing the transmitters from the aircraft with a like-new transmitter, it was determined that the left transmitter in the accident aircraft was at the 2.20 EPR position. The transmitter was electrically operated to verify the position for a 2.20 EPR reading. This could not be accomplished because the synchros would not stabilize. The right EPR transmitter was checked. The synchros were found to be movable and could not be used to determine impact position.

The engine instruments were damaged slightly by impact and remained attached to their panels. The gauges indicated the following:

Indication	Engine No. 1 (Left)	Engine No. 2 (Right)
N_1	0%	78%
EPR	2.9, Bug at 2.02	2.98, Bug at 2.26
EGT	820	220
N_2	23 percent	0
Fuel flow	3800 lb/hr	1200 lb/hr
Oil pressure	72 psi	53 psi
Oil temperature	Off scale −40°C	150°C
Oil quantity	4 gallons	3.5 gallons

7.4.2.g Structures Group Report

Wreckage recovery was initiated immediately after the accident and simultaneously with the recovery of victims. Recovery operations were conducted in coordination with the National Transportation Safety Board by various segments of the Department of Defense, the Department of Transportation, and the Metropolitan Washington Police Department, all under the general direction of the Federal Emergency Management Agency (FEMA).

As the aircraft descended, the right wing was structurally damaged when it hit a boom truck (Figures 7.7 and 7.8). Shortly thereafter, the aircraft struck the steel barrier and railing on the west side of the 14th Street Bridge (Figure 7.9) at an elevation of about 37 feet mean sea level. Fragments of the right wing remained on the bridge. The remainder of the wreckage sank in the Potomac River in about 25 to 30 feet of water (Figures 7.10 through 7.13). The wreckage area was confined to the south side of the river between the 14th Street Bridge and the Center Highway Bridge of the George Mason Memorial Bridge (Figure 7.14).

Figure 7.7   Boom truck struck by aircraft on bridge. (Courtesy NTSB.)

After initial impact, the aircraft broke into several major pieces. The fuselage broke into four major pieces: (1) nose section with cockpit (Figure 7.15); (2) fuselage section between nose section and wing center section; (3) fuselage-towing intersection; and (4) aft body structure with empennage attached. The wing structure was separated into three major pieces: (1) left wing outboard of the No. 1 engine, including all associated flight control surfaces; (2) wing center section, lower surface, including wing lower surface stubs between the No. 1 engine mounts and the No. 2

Figure 7.8 Boom truck and other vehicles struck by aircraft on bridge. (Courtesy NTSB.)

Figure 7.9 Steel barrier and railing struck by aircraft on west side of bridge. (Courtesy NTSB.)

Figure 7.10 Closeup of aft fuselage and empennage during recovery from river. (Courtesy NTSB.)

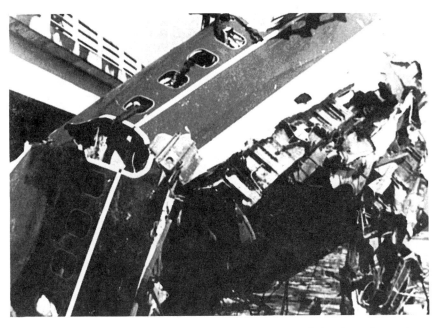

Figure 7.11 Recovery of center section of wing and fuselage from river. (Courtesy NTSB.)

Figure 7.12 Recovery of forward fuselage section from river. (Courtesy NTSB.)

Figure 7.13 Recovery of aft fuselage and empennage assembly from river. (Courtesy NTSB.)

engine mounts; and (3) right wing outboard of the No. 2 engine with the outboard 20 feet mostly disintegrated. The left main landing gear was separated from the wing, and the right main gear remained attached except for the wheels and oleo piston. The nose landing gear and its attaching structure were separated from the nose section. Both engines and their pylon structures were separated from the wings. There was no evidence of fire on any of the recovered structure.

Investigation of the flight controls and landing gear showed that the horizontal stabilizer jackscrew measured 7.75 inches between the upper stop and the traveling nut, a measurement that corresponds to 2.3 stabilizer leading-edge-up, or 5.3 units of trim. This stabilizer setting is in the green band for takeoff.

The B-737 aircraft has four leading edge flaps and six leading edge slats. Erosion was minimal on all of the leading edges and was within specified limits. Actuator extension for leading edge slats 1 and 2 was measured at $8\frac{1}{2}$ inches and 8 inches, respectively, which is consistent with trailing edge flaps 5 extended position. The actuator No. 3 leading edge slat was bent with about $8\frac{1}{4}$ inches of rod extension, which is consistent with the trailing edge flaps 5 position. The aileron trim assembly (in the wheel well) was found with the rig pinholes aligned. This corresponds to a zero trim setting for the ailerons. All segments of the trailing edge flaps sustained varying degrees of impact damage. Measurements taken between the travel nut and plastic cap at the trailing end of the flap jackscrews corresponded to trailing edge flaps 5 extension. All spoiler actuators were in the retracted (down) and locked position.

The right elevator and its trim tab were intact and were attached to the right stabilizer at all hinge points. The left elevator remained relatively intact and attached to the stabilizer. The rudder remained intact and attached to the vertical stabilizer, but no valid rudder trim measurements could be obtained.

Because of the extensive fragmentation, the integrity of the flight control system before impact could not be determined. Nearly all bellcranks, sector pulleys, and other mechanisms were broken, distorted, and separated from their attachment structures.

The nose landing gear was separated from its fuselage attachment structure. The nose gear strut and tires remained intact. There was no visible damage to the tires, and both wheels were free to rotate. The nose gear actuator was attached to the nose gear assembly, and the actuator was in the extended position.

The left and right main landing gear were torn loose from the aircraft. The right gear was recovered as a unit; the left gear oleo strut, piston, brakes, wheels, and tires were not recovered. Uplock mechanisms were undamaged and were in a configuration consistent with a gear-extended position.

The wing leading edge slat skin was subjected to a metallurgical examination. The *Boeing 737 Structural Repair Manual* (57-50-3), dated August 1, 1981, under the heading Wing Leading Edge Slat-Skin Erosion Repair, states: "Flight operation is not allowed if leading edge roughness is equivalent to or greater than that of 240-grit sandpaper." Although this operations limitation is not in the FAA-approved flight manual, it is a guide to inspection and maintenance personnel as to the conditions under which an aircraft should be released for flight operations.

To determine the surface roughness of representative sections of leading edge slat skin, specimens were taken from each of the slats of the aircraft. Samples of 240-, 320-, 400-, and 600-grit sandpaper were compared with each of the six skin sections. The comparisons were made by rubbing each of the surfaces with the fingertips, applying the same relative pressure. Numerous laboratory personnel performed the test. In all cases, the six slat skin specimens were found to be smoother than 600-grit sandpaper.

A sample of the slat leading edge skin surface and samples of 240 and 600-grit sandpaper were examined with the aid of a scanning electron microscope. Results of this examination also disclosed that the slat skin leading edge outer skin surface was smoother than 600-grit sandpaper.

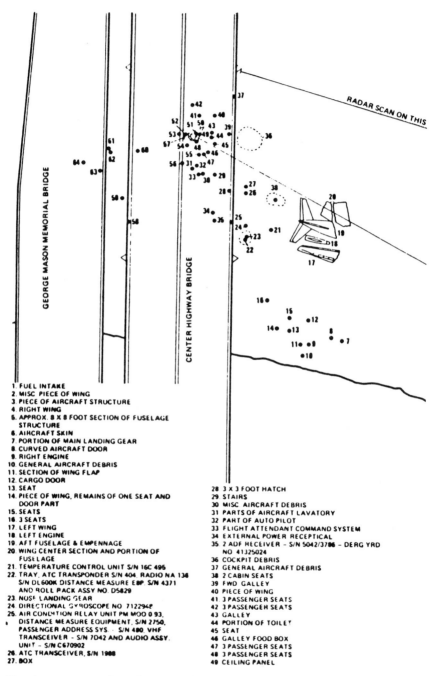

1. FUEL INTAKE
2. MISC PIECE OF WING
3. PIECE OF AIRCRAFT STRUCTURE
4. RIGHT WING
5. APPROX. 8 X 8 FOOT SECTION OF FUSELAGE STRUCTURE
6. AIRCRAFT SKIN
7. PORTION OF MAIN LANDING GEAR
8. CURVED AIRCRAFT DOOR
9. RIGHT ENGINE
10. GENERAL AIRCRAFT DEBRIS
11. SECTION OF WING FLAP
12. CARGO DOOR
13. SEAT
14. PIECE OF WING, REMAINS OF ONE SEAT AND DOOR PART
15. SEATS
16. 3 SEATS
17. LEFT WING
18. LEFT ENGINE
19. AFT FUSELAGE & EMPENNAGE
20. WING CENTER SECTION AND PORTION OF FUSELAGE
21. TEMPERATURE CONTROL UNIT S/N 16C 496
22. TRAY, ATC TRANSPONDER S/N 404, RADIO NA 134 S/N DL600K DISTANCE MEASURE EBP. S/N 4371 AND ROLL PACK ASSY NO. D5829
23. NOSE LANDING GEAR
24. DIRECTIONAL GYROSCOPE NO 7122942
25. AIR CONDITION RELAY UNIT PM MOO 0 93, DISTANCE MEASURE EQUIPMENT, S/N 2750, PASSENGER ADDRESS SYS. S/N 480, VHF TRANSCEIVER - S/N 7042 AND AUDIO ASSY. UNIT - S/N C670902
26. ATC TRANSCEIVER, S/N 1988
27. BOX

28. 3 X 3 FOOT HATCH
29. STAIRS
30. MISC AIRCRAFT DEBRIS
31. PARTS OF AIRCRAFT LAVATORY
32. PART OF AUTO PILOT
33. FLIGHT ATTENDANT COMMAND SYSTEM
34. EXTERNAL POWER RECEPTICAL
35. 2 ADF RECEIVER - S/N 5042/3786 - DERG YRD NO 41325024
36. COCKPIT DEBRIS
37. GENERAL AIRCRAFT DEBRIS
38. 2 CABIN SEATS
39. FWD GALLEY
40. PIECE OF WING
41. 3 PASSENGER SEATS
42. 3 PASSENGER SEATS
43. GALLEY
44. PORTION OF TOILET
45. SEAT
46. GALLEY FOOD BOX
47. 3 PASSENGER SEATS
48. 3 PASSENGER SEATS
49. CEILING PANEL

Figure 7.14 Wreckage location chart, reconstructed from underwater radar scan and manual plotting.

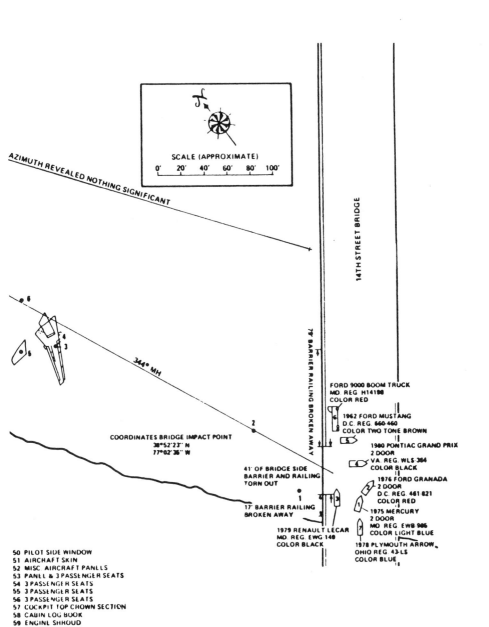

SCALE (APPROXIMATE)
0'   20'   40'   60'   80'   100'

AZIMUTH REVEALED NOTHING SIGNIFICANT

14TH STREET BRIDGE

344° MH

79' BARRIER RAILING BROKEN AWAY

FORD 9000 BOOM TRUCK
MD. REG. H14198
COLOR RED

1962 FORD MUSTANG
D.C. REG. 660-460
COLOR TWO TONE BROWN

COORDINATES BRIDGE IMPACT POINT
38°52'23" N
77°02'36" W

1980 PONTIAC GRAND PRIX
2 DOOR
VA. REG. WLS-364
COLOR BLACK

41' OF BRIDGE SIDE
BARRIER AND RAILING
TORN OUT

1976 FORD GRANADA
2 DOOR
D.C. REG. 461-821
COLOR RED

17' BARRIER RAILING
BROKEN AWAY

1975 MERCURY
2 DOOR
MD. REG. EWB-986
COLOR LIGHT BLUE

1979 RENAULT LECAR
MD. REG. EWG 140
COLOR BLACK

1978 PLYMOUTH ARROW
OHIO REG. 43-LS
COLOR BLUE

50 PILOT SIDE WINDOW
51 AIRCRAFT SKIN
52 MISC AIRCRAFT PANELS
53 PANEL & 3 PASSENGER SEATS
54 3 PASSENGER SEATS
55 3 PASSENGER SEATS
56 3 PASSENGER SEATS
57 COCKPIT TOP CROWN SECTION
58 CABIN LOG BOOK
59 ENGINE SHROUD
60 3 PASSENGER SEATS
61 3 PASSENGER SEATS
62 3 PASSENGER SEATS
63 3 PASSENGER SEATS
64 SEAT BACK

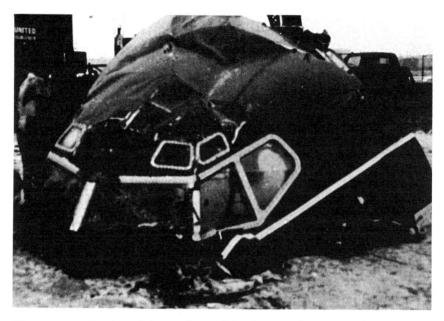

Figure 7.15 Section of forward top of fuselage (cockpit area) after recovery. (Courtesy NTSB.)

7.4.2.h *Systems Group Report*

The captain's and first officer's altimeters and vertical speed indicators were recovered from the river (Figure 7.16(c)). As Pitot-static system covers had not been used during deicing operations at the gate, the Safety Board sought to determine if deicing fluid had been introduced into the system, and submitted these instruments to the Federal Bureau of Investigation (FBI) laboratory for analysis. The FBI analysis revealed no trace of glycol. However, because of the prolonged immersion of these instruments in water after the crash, the results of the tests are not to be considered conclusive evidence that deicing fluid was either introduced or not introduced into the Pitot-static system.

A ground proximity warning system (GPWS) was installed on the aircraft. Mode 3 of the five-mode system indicates altitude loss after takeoff or go-around. The system is armed when the radio altimeter senses 100 feet, and it will sound an alarm when the barometric altitude loss is as little as 15 feet. However, if the aircraft never reaches 100 feet or never has a barometric altitude loss of at least 15 feet, the GPWS will not sense and, therefore, will not give an aural or visual warning. There is no evidence that the GPWS activated at any time during the flight of the accident aircraft.

7.4.2.i *Survival Factors Group Report*

At 1603 on January 13, the duty officer at the Washington National Airport fire station dispatched CFR equipment based on an intercepted radio transmission between the Washington National Airport Tower controller and the airport operations officer. While he was alerting the CFR crews, the crash phone rang at 1604, reporting the loss of visual and voice communication with an aircraft. The assistant fire chief on duty directed two CFR vehicles to respond to the end of runway 36 and directed three CFR vehicles to respond north on the George Washington Parkway beyond the airport boundary.

The assistant fire chief set up a command post on the shore of the river near the accident site at 1611. The fire chief arrived on the scene and assumed command of the crash site at 1620. At 1622, the airport airboat was launched. The boat launching ramp was covered with ice and the boat was literally picked up, moved to the frozen river, and launched.

In addition to the Washington National Airport CFR equipment, District of Columbia, Arlington County, Fairfax County, and the City of Alexandria fire departments responded.

For the six occupants who escaped from the aircraft, the temperature of both water and air was the major factor affecting their survivability. Water temperature 4 feet below the surface was 34°F. The survivors were in the icy water from 22 to 35 minutes before rescue. Survival time noted on the *Survival in Cold Water* chart[27] shows that, based on the water temperature, at least 50 percent of the survivors should have lost consciousness during that time period. All five survivors reported that the cold was so intense that they quickly lost most of the effective use of their hands; however, none reported loss of consciousness.

All but one of the survivors managed to cling to pieces of the floating wreckage. The one exception was the most seriously injured passenger. She was kept afloat by a life vest that was inflated and passed to her and her traveling companion by the surviving flight attendant. Her companion helped her to don the life vest. The survivors were unable to retrieve other life vests that were seen floating in the area. They reported that they experienced extreme difficulty in opening the package containing the one life vest that was retrieved. They stated that they finally opened the plastic package containing the life vest by chewing and tearing at it with their teeth.

Between 1622 and 1635, a U.S. Park Police helicopter rescued four passengers and one crew member and ferried them to the shoreline. When the rescue helicopter arrived, three of the survivors were still able to function sufficiently to help get themselves into the life ring and/or the

---

[27] National Aeronautics and Space Administration (NASA), *Bioastronautics Data Book*, SP-3006, p. 121.

a   b

loop in the rescue rope that was dropped by the helicopter crew. The other two survivors required hands-on rescue; one was pulled aboard the helicopter skid by the helicopter crewman, the other was rescued by a civilian bystander who swam out and pulled her ashore.

Three passengers from the aircraft, as well as two persons who participated in the rescue efforts, were placed in an ambulance and treated on the scene by paramedical personnel for hypothermia and shock. Radio communications were established with National Orthopedic Hospital and Rehabilitation Center in Arlington, Virginia, about 2 miles from the crash site. After treatment on the scene, the survivors were transported to National Orthopedic Hospital by ambulance.

Three factors are commonly used to determine survivability in an

C

Figure 7.16 (a) Simulated NORMAL engine instrument readings. Note normal N_1, exhaust gas temperature (EGT), N_2, and fuel flow. (b) Simulated ABNORMAL engine instrument readings. Note abnormal low N_1, EGT, N_2, and fuel flow. (c) Section of cockpit flight instrument panel on copilot side in accident aircraft. (Courtesy NTSB.)

aircraft crash (1) the decelerative forces do not exceed the tolerable limits of the human body; (2) the restraint system (seatbelts, seat structure, and seat anchorage points) remains intact; and (3) the occupiable area remains relatively intact to prevent ejection and provide living space for the occupants.

The primary impact forces experienced by the survivors did not exceed the tolerable limits of the human body. However, the secondary impact forces that most occupants experienced as a result of restraint system failures and violation of occupiable area did exceed these limits.

The recovered wreckage showed that the cabin separated from the cockpit and broke into three large sections and many smaller pieces. Virtually none of the cabin floor remained intact. All of the seats, whether empty or occupied, were extensively damaged and most were separated from the floor. The only occupiable space in the aircraft that remained intact and not violated by the collapsing cabin structure and furnishings was the area in the rear of the cabin, in the vicinity of the aft flight attendant seat.

7.4.2.j *Cockpit Voice Recorder and Flight Data Recorder Group Report*

A total of 82 divers trained to dive in icy waters were brought from various U.S. Navy, U.S. Army, and U.S. Coast Guard units to conduct salvage and rescue operations. Divers searched for the FDR and CVR using an acoustic device to home in on the discrete signals emitted by these recorders. Underwater visibility was 8 inches. Both recorders were recovered from the Potomac River on January 20, seven days after the accident.

The recorders were only superficially damaged. The foil recording medium was removed from the FDR, and examination disclosed that all parameters and binary traces were present and active. The altitude and airspeed traces were derived from the aircraft central air data computer. Other data recorded were magnetic heading, vertical acceleration, and radio transmitter (microphone) keying, all as a function of time (Figure 7.17).

A timing discrepancy was found in the FDR that made it necessary to evaluate carefully all values obtained from this unit. The first two radio transmissions were timed correctly, consistent with timing obtained from the CVR and ATC transcript within 1 second. The third transmission came only a measured 3 seconds after takeoff acknowledgment, rather than the 6 seconds indicated by the other sources. This discrepancy affected all recorded traces simultaneously and probably occurred a number of times throughout the accident flight. Examination of data from previous flights showed that it was irregular in occurrence and duration; the foil slowed for short periods of time then speeded up, rendering the overall timing correct while leaving short-term timing errors. This was caused by a malfunction of the foil takeup drive system. Therefore, FDR data were considered reliable only if validated by the other two data sources.

The FDR showed a sharp decrease and then a gentle rise in the altitude trace beginning at 1600:31.6 (0002:00.6); this is characteristic of the change in static pressure caused by aircraft rotation. Airspeed at this time was recorded as 130 knots. The V_2 callout occurred at 1600:37, and the sound of the stall warning (stickshaker) began 2 seconds later and continued until impact.

After rotation, the aircraft began to climb at a fairly constant but slightly decreasing airspeed; between 1600:37.6 and 1600:46.0, airspeed decreased from 147 to 144 knots. Altitude at the end of this period was 240 feet and heading had changed about 3° to the right. During the next 7 seconds ending at 1600:53.8, airspeed decreased significantly, from 144 to 130 knots, while heading changed to the left, from 009° to 002.4°; the maximum recorded altitude of 353 feet was achieved at the end of this period. The heading then continued changing to the left, reaching 347.5° 6.6 seconds later. The recording ended 0.6 second beyond this point with a heading of 354.4°.

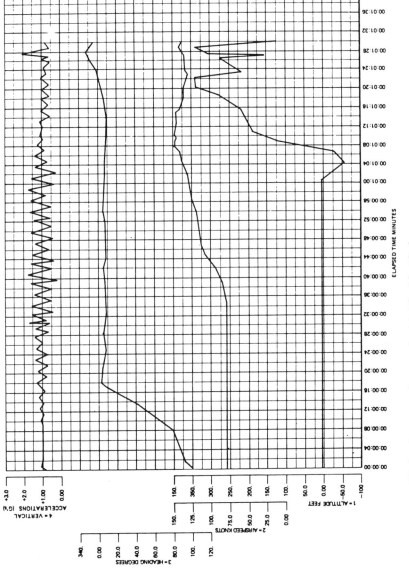

Figure 7.17 Flight data recorder graph, Air Florida Flight 90.

The altitude trace beyond 1600:54.0 is irregular, with rapid excursions up and down. The FDR altitude stylus was calibrated so that a movement of 0.0033 inch corresponds to a change in altitude of 100 feet (between 1000 and 8000 feet); hence, any vibration-induced stylus movement, such as might be experienced by stall buffet, would produce significant changes in the altitude trace with respect to the maximum value of 352 feet.

The CVR tape quality was good. As no timing signal is recorded on a CVR tape, timing was accomplished by adjusting the tape speed so that the 400-Hz aircraft power signal, which leaks onto the area microphone channel, was of the correct frequency. Copy tapes were made with a standard encoded time signal recorded on one channel. A timed tape was then compared with the tower tapes. (Tower tapes are recorded with a standard time reference signal from WWV—a radio station operated by the National Bureau of Standards.) Timing of CVR data is accurate to WWV time within 1 second.

A timed transcript of the cockpit area microphone channel and of radio communications data from the CVR was made. During preparation of the CVR transcript, members of the CVR group could not agree on the response to the checklist callout "anti-ice." The majority believed that the response was "off," but that word was put in parentheses in the transcript to indicate questionable text. The FBI Audio Laboratory was requested to perform an independent examination of that portion of the tape. The FBI concluded that the response to the checklist callout "anti-ice" was "off."

Events as recorded on the CVR and FDR were compared, and an overall matchup of the data from these two sources was compiled.

In this particular accident investigation, the CVR was also used to determine engine speed during the takeoff roll and flight by performing a "sound spectrum analysis." The CVR cockpit area microphone channel records sounds that originate or can be heard in the cockpit. In past accident investigations, particularly those involving aircraft with wing-mounted engines, the Safety Board had documented the engine sounds recorded on the cockpit area microphone channel. Experience and tests have shown that the predominant frequencies recorded are associated with the first and second stages of the low-pressure compressor fan blades of turbojet and turbofan engines. These frequencies are related to the rotational velocity of the fan by the number of blades in the first and second stages. Validity of the equations used to convert the sound spectrum analysis of blade frequencies into percent rotor speed was verified by Board investigators during tests at Boeing Aircraft Company, Seattle, Washington, on January 29, 1982.

For the spectrum analysis, signals from the cockpit area microphone channel were processed in a spectrum analyzer which displayed the energy content of the signals as a function of their frequencies. A number of these displays were printed to give a time history of the spectral content of the recording.

7.4.2.k *Performance Group Report*

Many issues related to aircraft performance were addressed by the performance group during the investigation of this accident. Only one of these issues, the degraded performance of the engines with a Pt_2 pressure probe blocked, is discussed in detail, as extracted from the NTSB report. Other issues, such as airframe icing, pitch up, and aerodynamic roughness, are covered in detail in the NTSB report.

The *B-737 FAA-Approved Flight Manual* and the *Air Florida B-737 Operations Manual* prescribe that the engine inlet anti-ice system shall be on when icing conditions exist, or when icing conditions are anticipated during takeoff and initial climb. The flight manual defines icing conditions as follows:

> Icing may develop when the following conditions occur simultaneously:
>
> The dry-bulb temperature is below 8°C (46.4°F). The wet-bulb temperature is below 4°C (39.2°F). Visible moisture, such as fog, rain, or wet snow is present.
>
> Fog is considered visible moisture when it limits visibility to one mile or less. Snow is wet snow when the ambient temperature is -1°C (30°F) or above.

The EPR measurement system in the B-737 aircraft senses an air pressure measured at the engine air inlet nose probe, known as Pt_2, and sets up a ratio between inlet air pressure and engine exhaust pressure measured at the engine exhaust nozzle, Pt_7. The EPR (Pt_7/Pt_2) is determined electronically and is displayed continuously in the cockpit. It is the primary instrument used by the crew to set engine power for takeoff. The Pt_2 probe is subject to icing but may be deiced with the engine anti-icing system. When the engine anti-icing system is manually activated by the crew, engine 8th stage compressor bleed air is supplied to the engine inlet guide vanes and is discharged into the engine nose cone and to the engine inlet upstream of the inlet guide vanes. This hot air keeps ice from forming or melts ice on the inlet probe by passing warm air around the probe, which is mounted in the nose cone.

With the engine operating, a false indication of the actual EPR can be indicated in the cockpit when ice blocks the inlet probe. Under this condition, the Pt_2 sensor will indicate a lower pressure, which leads to an EPR higher than the true value. Tests have demonstrated that with a blocked probe at takeoff, engine power can indicate an EPR of about 2.04 with the engine actually operating at an EPR of 1.70. In these circumstances, a pilot would unknowingly attempt takeoff at a considerably lower thrust than desired. However, the pilot has available other indications of engine operation displayed in the cockpit, such as a lower N_1, N_2, EGT and F/F consistent with the reduced engine thrust (Figures 7.16(a) and (b)).

Should the pilot activate the engine inlet anti-icing system with a blocked probe, the pilot would immediately notice a substantial drop in the indicated EPR, incorrectly indicating a low engine thrust as long as the normal Pt_2 sensing port remains blocked. This results from the introduction of engine anti-icing air flow into the nose cone and the resultant increase in pressure in the interior of the nose cone. This pressure is higher than would be sensed at the normal Pt_2 port. Falsely low indications of EPR have been detected by pilots when they found that they were unable to set takeoff power without exceeding redline N_1, N_2, and EGT.

7.4.2.1 *Human Performance Report*

At the time of the Air Florida accident, human performance investigation technology and NTSB staffing for this function were not as well developed as today. Human performance issues clearly existed in the area of cockpit resource management, particularly with regard to assertiveness by the first officer (copilot) during the takeoff phase of the flight. The following is the factual portion of the NTSB report covering human performance.

Air Florida pilots stated that their relationship with management is good, and that there is no pressure from management to keep to schedules in disregard of safety or other considerations. Current company statistics show that the upgrading period from first officer to captain averages about 2 years.

Three series of B-737 aircraft are flown by Air Florida pilots: the -100 basic, the -200 basic, and the -200 advanced. The accident aircraft was a -200 basic series B-737. There are some differences among these aircraft. In the -200 basic, there is a difference in the placement of engine instruments: the N_1 and EPR gauges are in reversed positions. The N_1 gauges are at the top of the engine instrument panel. Pilots indicated that they had not experienced any transition problems between the different aircraft types.

B-737 pilots told Safety Board investigators that they had not experienced any problems reading or interpreting the instrument displays or reaching or manipulating the controls. The NASA Aviation Safety Reporting System indicated that it had received no incident reports regarding crew station design in the B-737 aircraft.

7.4.2.m *Maintenance Records Group Report*

The aircraft was a Boeing 737-222. The aircraft, U.S. Registry N62AF, Serial No. 19556, was obtained by Air Florida from United Airlines on July 28, 1980, and had been operated continuously by Air Florida since that date. The aircraft total time as of January 13, 1982, was 23,608.44 hours at departure from Miami.

Statistical Data (Aircraft)
Date of certification: February 25, 1969
Fuselage number: P2600
Serial number: 19556

Registration number: N62AF
Airframe hours: 23,610:40 at departure DCA
Airframe cycles: 29,549 at departure DCA
An original standard Airworthiness Certificate, Transport Category, was issued February 20, 1969.
The aircraft was issued a valid Certificate of Registration dated October 2, 1980, in the name of Weiler, Alan G., Trustee, 1114 Avenue of the Americas, New York, NY 10036.
Statistical Data (Engines)
  Number 1 Engine
    Type: Pratt & Whitney JT8D-9A
    Date of manufacture: April 14, 1968
    Serial number: P655929B
    Total time: 20,762:20
    Time since overhaul: 20,762:20
    Total cycles: 26,955
    Date of installation: July 29, 1981
  Number 2 Engine
    Type: Pratt & Whitney
    Date of manufacture: July 2, 1971
    Serial number: P674546B
    Total time: 17,091:32
    Time since overhaul: 9171:15
    Total cycles: 16,661
    Date of installation: August 5, 1981

Under letter of agreement with American Airlines, the following services were provided to Air Florida, Inc., by American Airlines, Inc., at Washington National Airport:

Aircraft loading and unloading
Transport of passengers, baggage, mail, and cargo
Aircraft cabin cleaning
Weight and balance computation
Aircraft marshall and push out
On-call minor emergency aircraft maintenance as defined by American Airlines (Air Florida's technical representative to provide guidance and sign aircraft log)

7.4.3 Selected Findings and Conclusions

As a result of exhaustive investigation, tests and research, and a public hearing, which was held from March 1 until March 9, 1982, the NTSB made the following findings, determination of "probable cause," and recommendations.

7.4.3.a *Findings*

1. The aircraft was properly certified, equipped and maintained in accordance with existing regulations and approved procedures.
2. The flight crew was certified and qualified for the scheduled domestic passenger flight in accordance with existing regulations.
3. The weather before and at the time of the accident was characterized by subfreezing temperature and almost steady moderate-to-heavy snowfall with obscured visibility.
4. The aircraft was deiced by American Airlines personnel. The procedure used on the left side consisted of a single application of a heated ethylene glycol and water solution. No separate anti-icing overspray was applied. The right side was deiced using hot water and an anti-icing overspray of a heated ethylene glycol and water was applied. The procedures were not consistent with American Airlines' own procedures for the existing ambient temperature and were thus deficient.
5. The replacement of the nozzle on the Trump deicing vehicle with a nonstandard part resulted in the application of a less concentrated ethylene glycol solution than intended.
6. There is no information available in regard to the effectiveness of anti-icing procedures in protecting aircraft from icing which relates to time and environmental conditions.
7. Contrary to Air Florida procedures, neither engine inlet plugs nor pitot/static covers were installed during deicing of Flight 90.
8. Neither the Air Florida maintenance representative who should have been responsible for proper accomplishment of the deicing/anti-icing operation, nor the captain of Flight 90, who was responsible for assuring that the aircraft was free from snow or ice at dispatch, verified that the aircraft was free from snow or ice contamination before pushback and taxi.
9. Contrary to flight manual guidance, the flight crew used reverse thrust in an attempt to move the aircraft from the ramp. This resulted in blowing snow which might have adhered to the aircraft.
10. The flight was delayed awaiting clearance about 49 minutes between completion of the deicing/anti-icing operation and initiation of takeoff.
11. The flight crew did not use engine anti-ice during ground operation or takeoff.
12. The engine inlet probe (Pt_2) on both engines became blocked with ice before initiation of takeoff.
13. The flight crew was aware of the adherence of snow or ice to the wings while on the ground awaiting takeoff clearance.
14. The crew attempted to deice the aircraft by intentionally positioning the aircraft near the exhaust of the aircraft ahead in line. This was contrary to flight manual guidance and may have contributed to the adherence of ice on the wing leading edges and to the blocking of the engines' Pt_2 probes.
15. Flight 90 was cleared to taxi into position and hold and then cleared to take off without delay 29 seconds later.
16. The flight crew set takeoff thrust by reference to the EPR gauges to a target indication of 2.04 EPR, but the EPR gauges were erroneous because of the ice-blocked Pt_2 probes.
17. Engine thrust actually produced by each engine during take-

off was equivalent to an EPR of 1.70, about 3750 pounds net thrust per engine less than that which would be produced at the actual takeoff EPR of 2.04.

18. The first officer was aware of an anomaly in engine instrument readings or throttle position after thrust was set and during the takeoff roll.

19. Although the first officer expressed concern that something was "not right" to the captain four times during the takeoff, the captain took no action to reject the takeoff.

20. The aircraft accelerated at a lower-than-normal rate during takeoff, requiring 45 seconds and nearly 5400 feet of runway, 15 seconds and nearly 2000 feet more than normal, to reach liftoff speed.

21. The aircraft's lower-than-normal acceleration rate during takeoff was caused by the lower-than-normal engine thrust settings.

22. Snow and/or ice contamination on the wing leading edges produced a nose-up pitching moment as the aircraft was rotated for liftoff.

23. To counter the nose-up pitching moment and prevent immediate loss of control, an abnormal forward force on the control column was required.

24. The aircraft initially achieved a climb, but failed to accelerate after liftoff.

25. The aircraft's stall warning stickshaker activated almost immediately after liftoff and continued until impact.

26. The aircraft encountered stall buffet and descended to impact at a high angle of attack.

27. The aircraft could not sustain flight because of the combined effects of airframe snow or ice contamination which degraded lift and increased drag and the lower-than-normal thrust set by reference to the erroneous EPR indications. Either condition alone should not have prevented continued flight.

28. Continuation of flight should have been possible immediately after stickshaker activation if appropriate pitch control had been used and maximum available thrust had been added. While the flightcrew did add appropriate pitch control, they did not add thrust in time to prevent impact.

29. The local controller erred in judgment and violated ATC procedures when he cleared Flight 90 to take off ahead of arriving Eastern Flight 1451 with less than the required separation.

30. Eastern 1451 touched down on runway 36 before Flight 90 lifted off; the separation closed to less than 4000 feet, in violation of the 2-mile separation requirement in the Air Traffic Control Handbook.

31. Runway distance reference markers would have provided the flight crew invaluable assistance in evaluating the aircraft's acceleration rate and in making a go/no-go decision.

32. The Federal Aviation Adminstration's failure to implement adequate flow control and the inability to use gate-hold procedures at Washington National Airport resulted in extensive delays between completion of aircraft deicing operations and issuance of takeoff clearances.

33. The average impact loads on the passengers were within human tolerance. However, the accident was not survivable because the complex dynamics of impact caused the destruction of the fuselage and cabin floor which in turn caused loss

of occupant restraint. The survival of four passengers and one flight attendant was attributed to the relative integrity of the seating area where the tail section separated.

34. The crash/fire/rescue capability of Washington National Airport meets the applicable regulations, which do not require water rescue equipment.

35. Washington National Airport had water rescue equipment available; however, it had not been tested for use in ice-covered waters and it proved ineffective.

36. The Washington National Airport crash/fire/rescue personnel were notified 3 minutes after the crash as tower personnel attempted to determine the aircraft's whereabouts.

37. Rescue of the survivors was due solely to the expeditious response of a U.S. Park Police helicopter, and the heroic actions of the helicopter crew and of one bystander.

7.4.3.b Probable Cause

The NTSB determined that the probable cause of this accident was the flight crew's failure to use engine anti-ice during ground operation and takeoff, their decision to take off with snow/ice on the airfoil surfaces of the aircraft, and the captain's failure to reject the takeoff during the early stage when his attention was called to anomalous engine instrument readings. Contributing to the accident were the prolonged ground delay between deicing and the receipt of ATC takeoff clearance during which the airplane was exposed to continual precipitation, the known inherent pitchup characteristics of the B-737 aircraft when the leading edge is contaminated with even small amounts of snow or ice, and the limited experience of the flight crew in jet transport winter operations.

7.4.3.c Recommendations

As a result of this accident and several others involving operations in snow and icing conditions, the NTSB issued the following recommendations to the Federal Aviation Administration on January 28, 1982:

1. Immediately notify all air carrier operators of the potential hazard associated with engine inlet pressure probe icing, and require that they provide flight crews with information on how to recognize this hazard and require that flight crews cross-check all engine instruments during the application of takeoff power. (Class I, Urgent Action) (A-82-6)

2. Immediately review the predeparture deicing procedures used by all air carrier operators engaged in cold weather operations and the information provided to flight crews to emphasize the inability of deicing fluid to protect against reicing resulting from precipitation following deicing. (Class I, Urgent Action) (A-82-7)

3. Immediately review the information provided by air carrier operators to flight crews engaged in cold weather operations to ensure comprehensive coverage of all aspects of such operations, including the effects of a runway contaminated by

snow or slush on takeoff, and methods to be used to obtain maximum effectiveness of engine anti-ice during ground operations and takeoffs. (Class I, Urgent Action) (A-82-8)

4. Immediately require flight crews to visually inspect wing surfaces before takeoff if snow or freezing precipitation is in progress and the time elapsed since either deicing or the last confirmation that surfaces were clear exceeds 20 minutes to ensure compliance with 14 CFR 121.629(b) which prohibits takeoff if frost, snow or ice is adhering to the wings or control surfaces. (Class I, Urgent Action) (A-82-9)

5. Immediately issue a General Notice (GNOT) to all FAA tower and air carrier ground control personnel alerting them to the increased potential for aircraft icing during long delays before takeoff and when aircraft operate in proximity to each other during ground operations in inclement weather, and encouraging procedural changes where possible so that the controllers implement the gate-hold provisions of the Facilities Operations and Administration Manual 7210.3F, paragraph 1232. (Class I, Urgent Action) (A-82-10)

6. Document the effect of engine inlet pressure probe blockage on engine instrument readings and require that such information be added to approved aircraft flight manuals. (Class II, Priority Action) (A-82-11)

7. Amend Advisory Circulars 91-13c, "Cold Weather Operation of Aircraft," and 91-51, "Airplane Deice and Anti-ice Systems," to discuss in detail the effects and hazards associated with engine inlet pressure probe icing. (Class II, Priority Action) (A-82-12)

8. Revise the air traffic control procedures with respect to aircraft taxiing for takeoff, holding in line for takeoff and taking off to provide for increased ground separation between aircraft whenever freezing weather conditions and attendant aircraft icing problems exist. (Class II, Priority Action) (A-82-13)

9. Expand the training curricula for air traffic controllers and trainees to assure that instruction includes the hazards associated with structural and engine deicing of aircraft. (Class II, Priority Action) (A-82-14)

10. Immediately disseminate the contents of this safety recommendation letter to foreign operators involved in cold weather operations. (Class I, Urgent Action) (A-82-15)

7.4.3.d Response from the FAA to NTSB Recommendations

The National Transportation Safety Board has received the following response from the Federal Aviation Administration regarding the foregoing recommendations:

1. (A-82-6) On March 11, 1982, the FAA issued Air Carrier Operations Bulletin (ACOB) No. 7-82-2, Cold Weather Procedures, emphasizing the problems associated with engine inlet icing and suggested operational procedures.

2. (A-82-7, A-82-8) These safety recommendations were provided to all air carriers via a telegraphic message on January 29, 1982. A telephone conference between FAA's Office of Flight Operations personnel and all regional Flight Standards Division

Chiefs on January 29, 1982, tasked the regions to conduct a review of their operators.

3. (A-82-9) Reference to a time such as 20 minutes since deicing or the last confirmation that the aircraft surfaces were clear is not considered in the best interest of flight safety. Under some atmospheric conditions ice may form in a much shorter period whether ground deicing has been performed or not. Flight crews must use the "clear aircraft" concept specified by current rules without regard to specific time intervals. . . . As a result of the Air Florida Flight 90, B-737 accident and the subject recommendation, the R&D effort has been accelerated. We do not anticipate that changes will be made to the existing clean aircraft concept. However, information resulting from R&D efforts is expected to emphasize improved procedures to assure that hazardous ice formation does not exist prior to takeoff.

4. (A-82-10) A copy of NTSB Recommendations A-82-6 through A-82-15 was sent in its entirety to all air traffic facilities in GNOT form on January 28, 1982. The provisions of FAA Facilities Operations and Administration Manual 7210.3F, paragraph 1232, Gate-hold procedures, adequately cover the handling of departure procedure delays. The GNOT of January 28 acts to remind facilities to review their application of these procedures.

5. (A-82-11, A-82-12) A new Advisory Circular (AC) is being developed which will include a complete discussion of the hazards of engine inlet icing; pressure probe icing and blockage; and methods a flight crew can use to recognize these conditions and properly use the engine anti-ice system. In addition, a detailed technical analysis is being undertaken in order to include specific engine instrument reading impacts, cross-check procedures and performance degradation parameters in this AC. Initial information from this study is being immediately disseminated to the field in the ACOB described in the response to Recommendation A-82-6. When completed, this AC will be forwarded to the Safety Board. Flight Manuals will be revised after the AC is completed, if such changes are deemed essential for flight safety.

6. (A-82-13) The following note was added to Handbook 7110.65C: "Aircraft taxiing behind jet aircraft in freezing conditions may experience aggravated engine and airframe icing. For planning purposes, be alert to pilot advisories that increased taxi intervals may be used."

7. (A-82-14) In the meteorological portion of Phase II in the basic air traffic training program, in-depth training is conducted to identify the forms of icing and its effects on aircraft performance. Additionally, the FAA will advise the present work force via the Air Traffic Service Bulletin on the hazard associated with structure and engine icing of aircraft. The Bulletin will be published in September or October 1982.

8. (A-82-15) A telegraphic message was transmitted on January 28 to all FAA facilities; U.S. air carriers; U.S. owners, operators, aircraft and engine manufacturers; foreign authorities of known airplane registration; and other interested groups. . . . A survey report has been provided by each FAA region. This report indicates that all air carriers have been contacted and made aware of the safety recommendations and hazards associated with icing. The results of this survey indicate there is a positive attitude on the part of the industry concerning these safety recommendations.

As a part of its investigation, on May 10, 1982, the Board requested from the FAA data pertaining to the issuance or contemplated issuance of (1) changes to the Boeing 737 FAA Approved Flight Manual; (2) service bulletins applicable to the Boeing 737 aircraft; (3) operations manual bulletins applicable to the Boeing 737 aircraft; (4) revisions to the Boeing 737 Operations Manual; and (5) revisions to the Boeing 737 Maintenance or Structural Repair manuals.

On July 27, 1982, the FAA replied that its Seattle, Washington, Area Aircraft Certification Office has requested that Boeing change appropriate airplane flight manuals to more adequately cover the questions raised concerning B-737 and B-727 airplane icing of the engine inlet total pressure probe (Pt_2) and the use of engine anti-ice while on the ground. Boeing has issued a Telex to all operators of B-737's and B-727's on the subject of engine Pt_2 icing. Boeing is reviewing both the operations and maintenance manuals.

7.4.3.e Further Recommendations

As a result of the NTSB analysis of this accident, on August 11, 1982, the Board issued 11 additional recommendations to the Federal Aviation Administration, along with the final investigation report. These additional recommendations will not be reviewed here, but are available in the public domain.

7.5 THE PURPOSE OF INVESTIGATION: PREVENTION OF ACCIDENTS

The foregoing case study illustrates that the principal goal of an accident investigation by the National Transportation Safety Board is the prevention of similar accidents in the future. Separation of the litigation or fault-finding aspects of the investigation from the process encourages a thorough and timely investigation, with opportunities for early recommendations to correct deficient procedures.

The Air Florida Potomac River accident investigation case study, with the related prompt formulation and adoption of specific safety recommendations, clearly demonstrates the principal objective and desired end result of transportation accident investigation. Several of these recommendations might have prevented the Air Florida disaster, and will certainly be instrumental in preventing a recurrence of such an accident.

In a speech before the First Annual Seminar of the International Society of Air Safety Investigators[28] in 1970, the Hon. John R. Reed, former

---

[28] The International Society of Air Safety Investigators (ISASI) was originally incorporated on August 31, 1964, as an international nonprofit organization dedicated to the advancement of air safety through the exchange of ideas, experiences, and information acquired in aircraft accident and incident investigation. The official motto of ISASI is "Safety through Investigation."

Governor of Maine, then Chairman of the NTSB and later U.S. Ambassador to Sri Lanka, summarized the function of the NTSB as follows:

> We have certain skills and attitudes which we believe place the Board in a vital role in the civil aviation safety system. It is accident investigation combined with special studies that will best tend to prevent accidents.

Ambassador Reed concluded his address with this quotation from a wise observer:

> To look on is one thing,
> To see what you look at is another,
> To understand what you see is a third,
> To learn from what you understand is still something else,
> But to act on what you learn is all that really matters.

Several years later the NTSB, in its publication "Investigation Manual—Aircraft Accidents and Incidents,"[29] formalized the goals and objectives of aircraft accident investigation in these terms:

1. To maintain objectivity at all times and assure that each investigation is conducted with orderly thoroughness so that proper assessment of the probable cause can be made.
2. To assure that every investigation is studied sufficiently to identify hazards for which practicable safety recommendations can be developed which, when effectively implemented, would promote safety in transportation.
3. To assure that all personnel have the perseverence, dedication and training essential to the successful conclusion of every investigation.
4. To assure that all available skills and facilities of both government and industry are used in each investigation to the extent necessary to fully develop the facts, conditions and circumstances and the underlying causes involved in each accident.
5. To produce high-quality reports in a timely manner.

The philosophy of accident investigation from an international perspective is well stated in the "Third Edition—Manual of Aircraft Accident Investigation—International Civil Aviation Organization":

> Accident investigation is recognized as one of the fundamental elements of any sound programme for the improvement of air safety and the prevention of accidents. The high quality of accident investigation that is necessary to make a success of such a programme can only be satisfactorily achieved by specifically appointed accident investigators who are specially trained to perform the task.

The basic methodology presented here can be applied effectively to all types of transportation accident investigation, as well as to the investigation of other system failures and malfunctions (Wise and Debons 1987). The concept is to utilize the best available technical expertise and

---

[29] NTSB Order 6200.1A 4/25/80.

independent and unbiased investigation leadership to collect all possible factual data, to analyze the data, to determine the probable causes and contributing factors, and, finally, to make viable recommendations to prevent future accidents.

References

Burnetti, A. W. 1985. Development of aircraft accident investigation. *Flight Safety Digest* (Flight Safety Foundation) (June).

Federal Aviation Administration. 1987. *Airman's Information Manual.* Washington, DC: U.S Department of Transportation, April 9.

National Transportation Safety Board (NTSB). 1980. *Investigation Manual—Aviation,* NTSB Order 6200.1A 4/25/80.

_____ 1982. *Accident File, NTSB AAR-82-8, Air Florida Inc. Boeing 737-222, N62AF, Washington, D.C., January 13, 1982.* August.

_____ 1985. *NTSB—Annual Report to Congress.*

_____ 1987a. *Aircraft Accident/Incident Investigation Procedures.* Title 49 Transportation—Chapter VIII Part 831 3/3/87.

_____ 1987b. *Rules of Practice: Accident/Incident Hearings and Reports,* Title 49 Transportation—Chapter VIII Part 845 3/3/87.

Slonena, S. 1980. *The Contribution of Accident Investigation to Air Transportation Safety.* McDonnell Douglas Corporation–Douglas Aircraft Company, Report MDC J2100, November 20.

Society of Air Safety Investigators (SASI). 1971. *Forum,* Spring Edition (April).

Wise, J. A., and A. Debons. eds. 1987. *Information Systems: Failure Analysis.* Westinghouse Research and Development Center, University of Pittsburgh, NATO ASI Series. Berlin: Springer-Verlag.

8. Civil Engineering Investigation

GLENN R. BELL, P.E.

8.1 INTRODUCTION

Although specialists have studied failures of constructed facilities for decades, vigorous interest in the subject has developed over the past several years, especially in the United States. One of the reasons is the rash of spectacular collapses that occurred in the late 1970s and early 1980s. Evidence of this expanded interest is seen in the following activities:

In 1982 the Architecture and Engineering Performance Information Center (AEPIC) was formed. (See Section 2.5.)

In 1982 the Committee on Forensic Engineering (CFE) of the American Society of Civil Engineers (ASCE) was formed. In 1984 the CFE became a technical council of ASCE. Today there are six committees of the Technical Council of Forensic Engineering (TCFE), addressing a broad scope of failure issues.

In 1983 the National Academy of Forensic Engineers (NAFE) of the National Society for Professional Engineers (NSPE) was formed. Today the NAFE has over 250 members and publishes a journal. (See Section 1.4.1.)

Since 1984 the ASCE has sponsored nationwide short courses on forensic engineering. Video- and audiotapes for professional education have been produced on various topics in civil forensic engineering.

This activity addresses two broad interests: (1) the recognized need of our profession to learn from its failures and (2) the rapidly increasing number of engineering professionals specializing in forensic engineering.

In examining the activities of the American Society of Civil Engineers, one sees that the scope of civil engineering investigations can be quite large, encompassing the following disciplines:

Structures

Geotechnical works

Highways

Waterway, port, coastal, and ocean facilities

Culverts and pipelines

Air transportation and aerospace

Environmental facilities

Hydraulics, irrigation, and drainage installations

This chapter emphasizes investigations in two of these disciplines, structures and geotechnical works, although most of the principles apply to the other disciplines as well. More detailed description of traffic accidents is contained in Chapter 6, of air transportation disasters in Chapter 7, and of environmental systems failures in Chapter 9.

Although, as described in Chapter 1, forensic engineers commonly are concerned with the engineering aspects of legal problems, the activities of civil forensic engineers are much broader, addressing all issues of failures of constructed facilities, legal or not. This is demonstrated by the following stated purposes of the six committees of the ASCE/TCFE (ASCE 1988):

> *Task Committee on Guidelines for Failure Investigation.* To develop and publish guidelines to help the forensic engineer understand the process of failure investigation. These guidelines include all phases of the process from initial site visit, through report preparation to adjudication.
>
> *Publications Committee.* To review manuscripts, technical notes, and discussions submitted to the Council for publication in the *Journal of Performance of Constructed Facilities* and to be responsible for the publication of papers sponsored by the Council.
>
> *Education Committee.* To find ways to improve the preparation of civil engineering students for professional practice by means of failure case history utilization.
>
> *Committee on Dissemination of Failure Information (CDFI).* To develop and implement means for disseminating accurate and complete information regarding the nature and causes of structural failures in civil-engineered projects. CDFI will also act as a vehicle for coordinating and assisting in the dissemination of forensic engineering information generated by other organizations, both within and from outside ASCE. CDFI will act in an advisory capacity to AEPIC efforts in information dissemination and as a primary liaison between AEPIC and ASCE.

Committee on Practices to Reduce Failures. To develop objectives for change in the design and construction industry specifically targeted to reducing the incidence and severity of failures in civil-engineered projects; to pursue implementation of these changes and to act as liaison between ASCE and other professional societies and organizations to pursue the prevention of failures.

Research Committee. The task of the committee is to identify research needs related to the investigation of constructed facilities that do not perform in accordance with the established standards; define the research problems, and assign priorities to them; develop effective means of implementing the results of research.

Figure 8.1 The historic Hotel Vendome in Boston, Massachusetts, collapsed suddenly during a fire.

The author of Chapter 2 states that forensic engineers deal with a broad definition of "failure"—not all failures are catastrophic collapses. This is indeed very true for civil engineering investigations. Catastrophic news-making failures, like those shown in Figures 8.1–8.3, are rare in civil engineering works; most civil engineering investigations deal with lesser failures, such as deteriorated parking garage slabs (Figure 8.4), deteriorated and bulging exterior walls (Figures 8.5 and 8.6), and cracked concrete elements (Figure 8.7). Some failures, like fatigue and fracture, require specialized expertise (Figure 8.8). The Architecture and Engineering Performance Information Center indicates that leaky roofs and walls are possibly the most pervasive problem in buildings today (Loss and Kennett 1987). Although these lesser failures do not cause injury or death, they can be very costly to repair, and the lessons we have to learn from them are just as significant as those from catastrophic failures.

Investigations are conducted for a variety of purposes and clients. Usually where a substantial sum of money is concerned, the case will be involved in litigation or some other form of dispute resolution, although only a small fraction of cases may actually go to trial. The litigation or dispute resolution is not the sole focus of the forensic engineer's activities, however; it is merely one phase of the assignment in which the engineer presents the findings of the investigation. In other cases, litigation may not be anticipated or initiated. An owner may want to determine what has gone wrong, why a failure has occurred, and how it can be repaired, without any intention of seeking compensation for loss.

Figure 8.2 Collapse of Brazos River Bridge, Texas.

Figure 8.3 Failure of a crane during erection destroyed this catalytic cracking tower.

Figure 8.4 Severe deterioration at underside of parking garage slab.

Figure 8.5 Concrete deterioration at exterior bearing wall.

Figure 8.6 Investigation of brick veneer wall bulging at shelf angles.

Figure 8.7 Cracking of columns in prestressed concrete garage, Lowell, Massachusetts.

Below is a brief list of the types of clients by whom a civil forensic engineer may be engaged:

Owners

Developers

Public and government agencies

Plaintiffs in litigation (injured parties)

Defendants in litigation (generally, anyone involved with the design, construction, maintenance, or operation of a constructed facility)

Tenants

Attorneys

Insurance companies

Materials manufacturers

Designers

Contractors

Figure 8.8 Axle and bearing failures were responsible for several derailments on the New York–New Haven commuter rail service.

This chapter deals broadly with forensic civil engineering. Its purpose is to acquaint the reader with the qualifications and activities of the forensic civil engineer in an investigation, and to demonstrate the scientific process of information gathering and analysis that is part of any investigation. As a result of the author's experience, the chapter concentrates on the structural and geotechnical disciplines. Related topics are covered in other chapters (Forensic Photogrammetry—Chapter 10, Report Writing—Chapter 11, and the Role of the Engineer in Litigation and Other Dispute Resolution—Chapters 12 and 13).

To the author's knowledge, no universities offer degrees in civil forensic engineering, although a handful offer an individual course. There is no published definitive procedure for conducting an investigation, nor will there ever be; each investigation is different and requires a process of data collection, formation of failure hypotheses, and analyses of these hypotheses (Crist and Stockbridge 1985). For the practitioner, several publications touch on the techniques required for a competent failure investigation. In the area of structural investigations, only one of these is comprehensive, *The Guide to Investigation of Structural Failures* (Janney 1986). The ASCE Technical Council on Forensic Engineering is producing a new document, *Guidelines for Failure Investigation*, expanded to cover the broadest interests of the civil forensic engineer. It is scheduled to be published in 1989.

8.2 QUALIFICATIONS OF THE INVESTIGATOR

Civil engineers are trained largely for the design and construction of new works. Few successful designers are good failure investigators; the reverse is also true. The handful of exceptions are talented individuals who have extraordinary insight into engineering principles; they can apply that insight to either function. As designers, they often are responsible for innovative or daring designs. A good designer has

Familiarity with building codes, specifications, and industry standards

An understanding of accepted simplistic models of structural and soils behavior that are applied in design

Good managerial and organizational skills that lead to error-free designs at minimum design cost

A creative mind to develop simple and efficient solutions (particularly in details)

A good failure investigator, on the other hand, has

An ability to go beyond building codes and specifications and simplistic models of behavior to understand how facilities really behave and why they fail

An interdisciplinary understanding of constructed facility systems

An analytical and objective mind that can collect the needed data, develop failure hypotheses, and scrutinize those hypotheses to reach a correct conclusion regarding the causes of the failure

The key to proficiency in either of these areas is experience. Thus, I would recommend to anyone interested in developing expertise in investigating civil engineering failures a thorough study of documented investigations. It is unfortunate, however, that such information is at present not easily obtained by the practicing engineer. A handful of books have been produced on the subject (Feld 1968, LePatner and Johnson 1982, Ropke 1982, Ross 1984); the depth of the cases reported in these, however, is variable. Professional magazines such as *Engineering News Record* and *Civil Engineering* cover failures; the reports therein are not detailed, but are useful. Occasionally, failures are reported in the technical journals, for example, the articles by Fairweather (1975), Schousboe (1976), Loomis et al. (1980), Smith and Epstein (1980), and Pfrang and Marshall (1982). In 1987 the ASCE, in cooperation with other professional organizations, began publication of the *Journal of Performance of Constructed Facilities*, addressing both individual case studies and generic issues in civil forensic engineering. The National Bureau of Standards has produced a limited number of excellent reports on its investigations of failures (NBS 1982a–c, 1987). For those who are so fortunate, access to the report files of a firm that investigates failures would be invaluable.

A developing comprehensive source of failure information is the Architecture and Engineering Performance Information Center. [Vannoy (1983) provides background information.]

Review of this type of information is valuable for two reasons. First, it provides further insight into the special techniques for failure investigation. Second, and more importantly, it reveals patterns of modes and causes of failures. It is an understanding of these patterns—how facilities really behave and why they fail—that separates the experienced investigator from the rest. Excellent compilations of the typical modes and causes of structural failures are given by Addleson (1982) and Janney (1986). The patterns are remarkably consistent. Flat-plate structures nearly always fail by punching shear. Precast concrete beams and slabs frequently fail at their bearings. A well-known failure investigator once said, "Catastrophic failures in buildings occur by instability or at locations where the load must turn a corner (i.e., connections)." This overgeneralization is not far off the mark. In addition to reports on failures, research test results of structural components or systems are also valuable; the manner in which the test configuration models behavior of a real facility must be recognized when interpreting these results, however. An attorney–client is an advocate; an engineering investigator must always be a finder-of-fact.

Desirable qualities of forensic engineers in general are covered in Chapter 1. These include the integrity and cognizance to conduct an impartial investigation. In the majority of cases, the investigator is retained by a client with special interests. Whether done consciously or not, these interests can color the investigator's thinking. The objective investigator must constantly be aware of this.

The potential for conflict of interest in civil engineering works is greater than for other disciplines because of the complexity of constructed facilities and the number of firms and individuals involved in their creation. The forensic engineer must decline any opportunity where prior association with the project or firms or individuals involved with the project may color or may reasonably be perceived to color the investigator's impartiality and objectivity.

Equally important is the investigator's ability to act as an expert witness, should testimony be required. Briefly, the required attributes include the necessary technical credentials, unwavering honesty and integrity, an ability to remain composed under hostile examination, and a capacity to convey simply and convincingly to laypersons opinions that sometimes involve complex technical issues. These are discussed in more detail in Chapters 1 and 12.

8.3 ACTIVITIES IN THE INVESTIGATIVE PROCESS

Although investigative assignments vary widely in type, scope, objective, and complexity, most require the fundamental activities shown in Table 8.1. The order in which these activities appear in this table is not intended to imply any sequence in time. Indeed, many activities are

Table 8.1 Common Steps in the Investigative Process

Commission of forensic engineer by client; definition of objective of investigation
Collection of background information; preliminary document review
Initial reconnaissance site visit; eyewitness interviews
Formulation of investigative plan; formation of project team
Comprehensive collection of documents; document review
Site investigation; sample collection
Theoretical analyses
Laboratory analyses
Development of failure hypotheses, analysis of data, synthesis of information, and formation of conclusions
Determination of procedural responsibilities for failures
Report writing

undertaken simultaneously. The investigative plan is continuously revised and refined. Failure hypotheses are analyzed, eliminated, added, and revised. Some investigators have developed elaborate flow charts for the investigative process, such as are shown in Blockley (1981) and Kaminetzky (1983). These can be valuable for certain specialized and complex assignments; it is impossible to draw detailed process charts applicable to all assignments, however.

The investigative assignment usually starts with a telephone call or letter from a client interested in retaining a forensic engineer. The client generally indicates the problem to be addressed, and the type of investigation and technical expertise required, although depending on the client's prior experience and sophistication he or she may not know exactly what and who is required. One of the first things you must establish is what and where the project is and who the key players are. Until you have established that you do not have a conflict of interest, you should caution the client not to reveal any information that could compromise his or her position, if you should find it necessary to decline the assignment.

Early activities will involve establishing with the client the scope and objective of your work, any operating restrictions that may be placed on your investigation, the terms of your compensation, your qualifications for the work, and the schedule. For large investigations this will likely involve an exchange of documents, a meeting with the client, and possibly a brief visit to the project.

To determine the causes of failure, the goal of these activities is to establish the following:

The mode and sequence of the failure

The demands acting on the facility at the time of failure

The capacity of certain components or of the entire facility at the time of failure

Establishing the mode and sequence of the failure may be a simple observation, or it may be complex, such as in the case of a catastrophic progressive collapse, where the initiating trigger is not readily apparent. The load effects may be static or dynamic, and they frequently involve environmental factors such as temperature and humidity. Establishing capacity includes determining the as-designed and as-constructed state of the facility, but it also may be necessary to establish the load and environmental history of the facility, such as in cases involving fatigue, fracture, or deterioration. It is highly desirable, but not always possible, to establish capacities by both analytical and physical testing techniques.

In this process the investigator seeks to determine the initiating location and mode of failure that is consistent with the presumed sequence of failure, and to ascertain that at that location, and for that mode of failure, the demands on the facility exceeded its capacity.

The activities of the investigative process outlined in this section are described in more detail in the following sections.

8.4   THE INVESTIGATIVE TEAM

On complex projects the principal investigator rarely has sufficient technical expertise to deal with all aspects of the investigation and must rely on experts of other disciplines; these experts may be within or outside of the principal investigator's firm. With continued technical specialization, the "interdisciplinary team" approach is increasing in popularity, even for small projects. A comprehensive, but by no means exhaustive, list of specialists who may be retained in civil engineering investigations is given in Table 8.2.

When the investigation is likely to involve litigation, these experts must be selected not only for their technical competence, but also for their capabilities as expert witnesses. The principal investigator should have a working relationship with such experts that will allow them to be called on short notice, in the event their services are needed on site.

On one large investigation in our office, four in-house group leaders of various disciplines and eleven outside individual experts or consulting firms reported to the principal investigator. In such cases, the principal investigator is involved solely with coordinating and analyzing the work of others. Communication among the various team members requires strong organization. Standard reporting procedures must be established. Periodic team meetings should be held, wherein the investigative plan is reviewed and, if necessary, refined, and failure hypotheses are discussed. The principal investigator with some appreciation for each of the specialized disciplines involved is a rare and valuable individual.

Examples of other team members who may not be considered experts in the sense implied above, but who are necessary in the team's field operations are given in Table 8.3.

Table 8.2 Specialist Consultants

Aerodynamics	Measurement technology
Aluminum materials	Meterology
Architecture	Ocean Engineering
Blasting vibrations	Offshore construction
Climatology	Painting
Cold weather construction	Parking engineering
Composite materials	Pavements
Computer design	Petrography
Concrete materials	Photoelasticity
Construction equipment and methods	Photogrammetry
Corrosion	Pile driving
Cost estimating	Pipelines
Data systems	Pipes
Dynamics	Plasticity
Elasticity	Plastic materials
Electronics	Prestressed concrete
Engineering mechanics	Probability theory
Environmental engineering	Protective coatings
Fabrication	Railroads
Fatigue	Shock
Field testing	Shoring
Fracture	Stability
Geology	Statistical analysis
Geotechnical engineering/foundation design	Steel materials
Glass	Surveying
Ground water	Timber
Hydraulics	Vibration
Hydrodynamics	Waterproofing
Impact	Welding
Masonry technology	Wind
Mathematics	

Table 8.3 Field Support Specialists

Surveyors
Concrete coring or sawing technicians (for sample removal)
Welders (for sample removal)
Photographers
Crane operators
Field and laboratory testing technicians
Measurement technicians
Witness interrogators

8.5 SITE INVESTIGATION AND SAMPLE COLLECTION

8.5.1 General

Whenever possible, the investigator should obtain and review construction drawings and other pertinent documents to generally become familiar with the facility before the initial site investigation. The principal investigator is responsible for establishing the location reference system (column or grid lines), member cataloging system, photographic numbering and cataloging system, and sample identification system for use by all team members.

Collapse or failure scenes can be dangerous places. Once, during an investigation of the collapse of an earth-covered garage, a second section of the garage collapsed (Kaminetzky 1976). Therefore, the leading concern of every team member must be his or her own safety and the safety of fellow team members.

In rescue and cleanup operations, evidence is altered or removed, possibly never to be available again. Thus, the speed with which the team acts can determine how much information is available in the investigation. This is where the experience of the investigator comes into play; he or she must formulate failure hypotheses, and direct observations and sample collection to those hypotheses. Often the site investigation team must make the best use of a limited amount of time. In these cases it is reasonable to emphasize the more likely hypotheses, but no avenue should be closed until it can be eliminated by positive proof.

Also when time is limited, it is legitimate often to request of the appropriate authority that the cleanup operation be altered or delayed in the interest of preserving data. This is provided, of course, that it does not impede any rescue operations where human well-being is involved.

The investigate team must understand clearly its authority in conducting its investigation: Does it have legitimate access to the site? To what extent can it direct alteration of evidence for its own benefit in observation? Does it have the right to remove specimens without consultation or knowledge of other potential investigators? A state or federal investigative team, when present, is usually the highest investigative authority on site.

When the team's access to the site is delayed, a review before the initial site visit of any available drawings and a general description of the failure help to focus the team's effort. Drawings are also valuable in guiding documentation of the as-built condition of the structure at the site. Do not neglect to verify, however, that the information shown on the drawings (plan dimensions, member sizes, and material properties, for example) represents the as-built state of the structure. An example of a building structure whose construction bore little or no resemblance to the construction documents is given by Bell and Parker (1987). It is easy to take things for granted.

Specific site activities are outlined below.

8.5.2 Equipment

Each team member should have his or her own equipment. A checklist of basic clothing, safety equipment, and hand tools is included in Table 8.4; this should suffice for most initial site investigations. Specialized equipment, such as for sample removal and in-place strength tests, may also be required, as discussed below.

8.5.3 Written Record of Observations

A written record of observations should constitute an entire log of the investigator's activities and should include the following:

Sketches of the overall failed configuration of the construction

Observation of behavior of adjacent construction during and subsequent to the failure

Detailed sketches of critical members and connections

Inventory of construction materials to establish dead loads

Observation of deterioration

Records of detailed as-built conditions, including plan and detail dimensions

Description of fracture surfaces

Records of samples removed

Indications of environmental conditions acting on the facility at the time of failure

A log of photographs

Records of conversations with others

Dimensional entries should also include the method of measurement used (e.g., dial caliper, steel tape, or other). Precision measurement devices should be calibrated periodically, records of calibration should be maintained, and measurement technicians should be trained in the use of such devices.

It seems trite to say that the neatness, organization, and completeness of the notes are important, but in failure investigations these things are especially important. The investigator may need to refer to notes years later, should the case be involved in litigation. The expert witness may need to refer to these notes in court or deposition, in which case the notes may become court exhibits, subject to close scrutiny by hostile parties. Some investigators prefer separate records for photograph logs, member sketches, and sample removal (Janney 1986). The author generally prefers

Table 8.4 Small Tools and Equipment for Field Investigation

Measurement	Stationery
10-inch dial caliper	White-lined paper
12-inch steel ruler	Calculation pad
20-foot tape	Architect scale
100-foot tape	Pencil and leads
Optical comparator	Eraser
Tape-on crack monitors	Field notebooks
	Triangles
Tools	Felt markers
Hammer	Clipboard
Screwdrivers	Lumber crayons
Prybar	
Pocket knife	Books
Flashlight	AISC manual
	ACI code
Photographic	Other applicable failure guides
Film	
Camera	
Lenses	Other
Flash	Calculator
Batteries	Stick-on labels
Lens papers	Wire-on labels
	Plastic sample bags
Clothing	Duct tape
Hardhat	Job file
Coveralls	Spray paint
Workboots	Dictation recorder
Gloves	Dictation tapes

to keep all records in one spiral-bound notebook, which keeps a running log of activities. Pages are less likely to be misplaced or lost in the often-rushed site activities, and the running log keeps track of the sequence of the investigator's activities; it is often important to know at what stage or in what context various activities took place.

Sketches usually provide more information in less space with less effort than text descriptions. For this reason, hand notes are usually superior to tape-recorded comments.

8.5.4 Photography

Photographs are an essential means of documentation. Some investigators use professional photographers. I prefer to take my own photographs, because I know what I want the photographs to show. There is no substitute for looking through the lens yourself. Most investigators should find worthwhile the financial investment in a good 35-mm SLR camera system and the time investment to learn basic photographic skills. A good cataloging trick is to make the first photo of each roll an identification shot of a note pad indicating the project number, date, roll number, and photographer. Portable video cameras in VHS or 8-mm format are useful when the qualities of sound and motion must be conveyed.

In collapse investigations overall aerial views, perhaps shot from a cherrypicker or an adjacent rooftop, help to record the overall mode or sequence of failure. Also take detailed shots of critical members, connections, and failure surfaces. For fracture surfaces, macro or stereo-image photographs are valuable. If the team arrives at the site during the rescue and cleanup operation, it should concentrate on general photographic coverage. Detailed shots can be saved for later.

Further information on photographic technology is given in Chapter 10.

8.5.5 Sample Removal

As mentioned earlier, the investigative team must understand clearly its legal right to remove samples, and it must consider needs of other investigators when selecting the number and location of samples to be removed. The inexperienced or overzealous investigator can quickly destroy valuable evidence. On projects that involve more than one investigative team, it is becoming increasingly common for all teams to cooperate and agree on the number, size, and location of samples to be taken and the methods of testing (Sowers 1987). The samples can be tested at an independent laboratory acceptable to all parties, and the costs and results can be shared. Sometimes the sharing of data, results, and associated costs is ordered by the courts.

Before removing any sample, label it and document its position and condition with notes and photographs (Figure 8.9). After removal, samples must be maintained by responsible parties under a chain of possession that is later traceable to the site.

To limit laboratory testing costs and minimize destruction of evidence, samples should be taken only to establish parameters significant to the investigation. Sample removal should be based on the team's failure hypotheses. The number and size of samples required may depend on the degree of certainty required in the information obtained and on the variation in observed data. This can be determined scientifically using a classical approach to estimation of parameters (Ang and Tang 1975). More frequently, the selection is based on more practical considerations, such as the availability of good samples, and on the judgment of the investigator.

8.5.6 Eyewitness Accounts

My experience with eyewitness accounts of failures is that they are often not accurate or reliable. This is not because eyewitnesses are dishonest, but because most do not have the necessary technical training, and when involved in catastrophes, they frequently are distracted by other issues and come away with a distorted view of the events. An exception to

Figure 8.9 Field samples must be carefully documented, labeled, and packed for shipping to the laboratory.

this is the information obtained from some construction workers, when the failure occurs during construction. The information obtained from an interview should be used to guide the investigator. Used with other corroborating interviews or information, the interview may provide useful evidence. The interview must be structured to elicit reliable information; leading questions must be avoided. The interviewer should record an assessment of the acuity of the witness, along with the record of statements. Janney (1986) presents detailed guidance for conducting the interview.

8.5.7 Field Tests—Structural Investigations

8.5.7.a *General*

The types of field tests used in structural investigations are too numerous to describe herein. Table 8.5 gives a list of reference standards of commonly used field tests; this list is by no means exhaustive, however. Brief descriptions of some of the general categories of tests are given below.

As with field sampling, the investigator must give careful consideration to the number of tests and the locations at which they are to be conducted. Some of the ASTM standards for individual tests give recommendations for such methodologies. Where no specific guidance is given, the investigator must rely on statistical analyses and/or judgment.

Table 8.5 Common Field Tests in Structural Investigations

Test	Reference (where applicable)
Load tests	
Full-scale load tests	Schreiver (1980), NAS (1969)
Cladding components	ASTM E997, E998
Beams and girders	ASTM E529
Floor and flat roofs	ASTM E196, E695
Truss assemblies	ASTM E73, E1080
General practice	ASTM E575
Instrumentation	
Electronic instrumentation and data acquisition for strain, deflection velocity, acceleration, pressure, force	Bendat and Piersol (1971), Colloms (1983), Jones and Chin (1983), and Smith (1985)
Dial gauge displacement measurement	
Tape and rule displacement measurement	
Optical or laser displacement measurement	
Taut-wire displacement measurement	
Dimensional measurements	
Conventional level survey	
Laser and infrared survey	
Liquid-filled tube	
Photogrammetry	
Satellite methods	
Concrete materials	
Windsor probe	ASTM C803
Swiss hammer	ASTM C805
Ultrasonic testing	
Tensile snap-off tests	
Anchor strength	ASTM E488
Metal materials	
Hardness testing—Brinell	ASTM A833, E10, A370, E140
Hardness testing—Rockwell	ASTM E18
Hardness testing—Vickers	ASTM E92
Hardness testing—Knoop	ASTM B528
Masonry materials	
Anchor pullout tests	ASTM E754, E488
Water permeance	ASTM E514
Weld testing	
Radiographic method	ANSI/AWS B1.10,
Ultrasonic method	ANSI/AWS D1.1,
Visual method	AWS (1980); ASTM E94, E142, E390,
Dye penetrant method	E1032, E164, E500, E273
Magnetic particle method	
Eddy current method	
Water and air penetration, heat loss, and air penetration through curtain walls (see also Table 8.9)	ASTM E283, AAMA A501.2, ASTM E1105
Subsurface investigations	
R-meter Radiographic	
X-ray Eddy current	
Pulse-echo	

When the test results are critical to the conclusions of the investigation, the test should be corroborated with results of similar tests of a different method or by calculations.

8.5.7.b *Load Tests*

The static, dynamic, and fatigue behavior under load of structural systems or components is frequently determined by testing in the field. Many types of loading are applied, including hydraulic jacks (Figure 8.10), mechanical jacks, air pressure, imposition of water or other heavy materials, reciprocating machinery, or impulse loading (NAS 1969; Schriever 1980). Before the test is conducted, it is important to isolate the component or region of the structure being tested from other structural and nonstructural components to accurately obtain its response. Safety precautions such as provision of limiting scaffolding that preclude complete collapse must be taken to avoid injury.

Detailed examination of the tested element (e.g., crack detection) should precede and follow the test.

8.5.7.c *Instrumentation*

The microcomputer revolution has produced a new dimension in field instrumentation capabilities for structures (Bendat and Piersol 1971; Colloms 1983; Jones and Chin 1983; Smith 1985). Common data-gathering

Figure 8.10 Load testing of large-diameter reinforced concrete pipe.

sensors include electronic strain gauges (Figure 8.11), electronic displacement transducers (linear potentiometers, rotary potentiometers, and LVDTs), accelerometers, and pressure transducers. Other displacement-measuring devices include dial gauges, simple tapes or rules, and optical or laser technology. Traceable calibration procedures must be strictly employed with all instruments. The operating limitations and performance characteristics of instruments must be clearly understood.

8.5.7.d *Dimensional Measurements*

Conventional level survey techniques are employed to measure building topography (as in the case of foundation settlements or overdeflection); laser and infrared technology, liquid-filled tubes with pressure transducers, high-resolution photogrammetry, and satellite methods have been refined to increase the accuracy and ease of data gathering, in some cases, over traditional methods. Equipment has been developed for specific dimensional applications, for example, for measuring the flatness of concrete floor slabs or the ovalling of buried pipe (Figure 8.12).

8.5.7.e *Concrete Materials*

Rebound methods, such as the Windsor probe or Swiss hammer, are available for estimating the *in situ* compressive strength of concrete. They are best used to indicate relative strengths of samples, after calibration with compressive strength results taken from cores for a particular project. Ultrasonic techniques have been used to determine concrete density, as well as to detect flaws such as cracks, as described later. Snap-off tests have recently been developed for estimating the tensile strength of concrete *in situ*.

8.5.7.f *Metal Materials*

Most metal testing is done by removing coupons for laboratory testing as described in Section 8.8. Hardness tests can be applied in the field to estimate metal tensile strength. Flaw detection is discussed under Subsurface Investigation.

8.5.7.g *Wood Materials*

Common wood tests applied in the field are simple methods that involve probing or boring into the surface of the wood to estimate its density or to detect deterioration such as rot, fungus, or insect infestation.

8.5.7.h *Weld Testing*

The techniques of testing welds of metal fabrications are standardized by the American Welding Society (AWS). Their application requires an inspection specialist. Common nondestructive testing methods employed include radiographic, ultrasonic, visual, dye penetrant, mag-

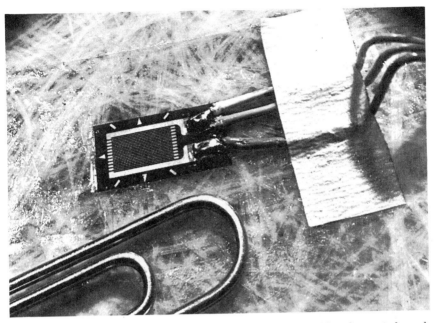

Figure 8.11 Electronic strain gauge used to instrument fiberglass-reinforced plastic.

Figure 8.12 Specially designed profilometer is used to measure deflected shape of buried reinforced plastic pipe.

netic particle, and eddy current methods. Selection of the best method for a particular task depends on the type of discontinuity to be detected, the joint type, and the accessibility of the joint (Betz 1986; AWS 1986).

8.5.7.i *Water and Air Penetration, Heat Loss*

These tests are most often used to check the capability of building cladding to provide a suitable enclosure from natural elements. Water and air penetration tests range from simple hose or flood testing to elaborate setups with spray racks and vacuum/pressure chambers. Methods are specified by the American Society for Testing and Materials (ASTM) and the Architectural Aluminum Manufacturers' Association (AAMA). Infrared thermography has recently been refined for general use in detecting heat loss; used correctly it can be an effective tool for varied applications (e.g., detection of air leakage in walls, presence of moisture in roofs, and location of reinforcement in concrete slabs).

8.5.7.j *Subsurface Investigation*

Various nondestructive methods are available to detect conditions below the surface of a material, such as flaws or embedments. Magnetic "R-meters," X-ray, and pulse-echo methods are used to detect reinforcing bars or other steel embedments in concrete and masonry. Radiographic, ultrasonic, and eddy current methods are used to detect subsurface flaws in metals.

8.5.8 Field Tests—Geotechnical Investigations

8.5.8.a *General*

The types of field tests used in geotechnical investigations are also varied, generally for determining the subsurface properties of soil and rock, groundwater hydrology, and the in-place performance of foundation and earth-retaining systems. Some common field tests are described here. A more detailed list of tests is given in Table 8.6.

8.5.8.b *Borings and Penetration Tests*

Used to sample soil and rock, evaluate strata locations, and estimate in-place strength or density, these methods include hand-auger drilling, power-auger drilling, hollow-stem-auger drilling, rotary drilling, diamond core drilling, auger sampling, split-barrel sampling, Shelby tube sampling, Giddings sampling, standard penetration tests, and dutch cone penetration tests.

8.5.8.c *Test Pits*

These are used to identify subsurface profile, obtain undisturbed samples, and conduct in-place strength tests.

Table 8.6 Common Field Tests in Geotechnical Investigations

Test	Reference (where applicable)
Boring and penetration tests	
Hand-auger drilling	ASTM D1452
Power-auger drilling	ASTM D1452
Hollow-stem-auger drilling	ASTM D1452
Rotary drilling	
Diamond core drilling	ASTM D2113
Auger sampling	
Shelby tube sampling	
Giddings sampling	
Standard penetration test	
Dutch core penetration test	
Test pits	
In-place strength tests	
Borehole shear test	
Vane shear test	ASTM D2573
Load tests	
Structural load test	
Pile load test	ASTM D3966, D1143
Rock bolt pull tests	ASTM D4435
Instrumentation	
Piezometers	ASTM D4448
Tiltmeters (see also Table 8.5)	ASTM D4622
Dimensional measurements	
(see Table 8.5)	
Density	
Drive cylinder test	ASTM D2937
Nuclear method	ASTM D2922
Sand cone test	ASTM D1556
Sleeve method test	ASTM E12, D4564
Seismic	ASTM D4428

8.5.8.d *In-Place Strength Tests*

Used to estimate soil shear strength in situ, these tests include the borehole shear test and the vane shear test.

8.5.8.e *Load Tests*

Foundation and earth-retaining structures are often load tested in place using techniques similar to those described earlier for structures. The most common is the pile load test, wherein a test pile is loaded by hydraulic jacks against a test frame.

8.5.8.f *Instrumentation*

Foundation and earth-retaining structures are instrumented using many of the same techniques described earlier for structures. Special devices developed for geotechnical instrumentation include piezometers and tiltmeters.

8.5.8.g *Dimensional Measurements*

Conventional and newly developed surveying techniques are used to detect foundation settlements and other soil movements. The techniques are described above for structures.

8.5.8.h *Seismic Tests*

Seismic techniques are used for rapid and broad determination of soil and rock strata.

8.6 DOCUMENT COLLECTION AND REVIEW

8.6.1 General

Document collection and review is usually an ongoing process throughout the investigation. At the outset of the assignment, the investigator attempts to obtain pertinent documents regarding the overall design and construction of the facility to become generally familiar with the project. However, not all needed documents may be available to the investigator immediately, nor is it necessarily immediately apparent to the investigator which documents are necessary.

The types of documents the investigator may use fall into two categories:

> Project-specific documents regarding the history of design, construction, modification, operation, and prior investigation of the facility in question
>
> Research documents regarding the characteristics and performance of key systems or elements of the facility

These categories are discussed separately.

On large, complex projects, document reviews can require enormous effort. The investigator must plan the review carefully and consider only those documents relevant to the topics at issue. On the largest projects of this type, our office has used database management computer programs to keep track of documents, to sort or retrieve information by subject, author, date, or source, and to perform statistical analyses and other data manipulations. Several programs are available to do this; two that we have used with success are *dBase III*, which runs on most personal computers (PCs), and *Easytrieve*, which runs on mainframes. In the use of PCs, large mass-storage units of tens or hundreds of megabytes capacity may be necessary.

8.6.2   Project-Specific Documents

Project-specific documents are reviewed to assist in determining (1) the operating condition of the facility (e.g., strength, serviceability, or process function) at the time of failure, (2) the operating effects acting on the facility (e.g., loads or environmental conditions) at the time of failure, and (3) the allocation of responsibilities of various parties for the causes of failure. The extent to which the investigator may be assigned to the third task can vary widely among projects.

Most documents are part of the project correspondence created during the facility's design and construction. In cases involving litigation, the investigator's access to these documents must be arranged through the client's attorney.

Occasionally, all parties in a legal dispute agree to make available to each other all documents in their files for the case or to turn the documents over to a neutral depository. More frequently, the client's attorney must request specific documents through the legal process. The attorney looks to the investigating engineer for assistance in this. A list of typical documents that the investigating engineer may request for a civil engineering investigation is included in Table 8.7. Not all of these may be appropriate for every project. Common sources of documents are given in Table 8.8.

Weather records of wind speed and direction, precipitation, temperature, and relative humidity are useful in establishing environmental conditions at the facility prior to and at the time of the failure. Common sources are the National Oceanic and Atmospheric Administration (NOAA), National Climatic Data Center in Asheville, North Carolina, and local weather bureaus.

Frequently in collapse investigations, the team may not be granted access to the site for some time after the collapse, perhaps after the debris has been removed. In such cases, information from others on the condition or configuration of the facility immediately after the collapse becomes vital. The best source is photographs. We have obtained such photos from newspapers, television news reporters, police departments, fire departments, civil defense rescue teams, and local building inspectors. Aerial photographs may be purchased from the U.S. Geological Survey (USGS) and other sources.

8.6.3   Research Documents

Normally the investigator researches documents that are not specific to the project to obtain information on analysis techniques or the characteristics and performance of key systems or elements of the facility. These documents may aid the investigator in determining the mode and technical cause of the failure. They may also be used in developing

Table 8.7 Project Specific Documents Used in Investigation

Contract drawings (including all revision issues thereof)
    Structural (including progress prints)
    Architectural (including progress prints)
    Mechanical (HVAC)[a]
    Electrical[a]
    Plumbing[a]
    Lighting[a]
Contract specifications
    Technical sections of interest
    General conditions
    Supplementary general conditions
    Special conditions
Contracts
    Owner/architect
    Architect/structural engineer
Contract revisions
    Addenda
    Bulletins
    Field directives
    Change orders
    Any correspondence authorizing change to the structure from the contract
      requirements
Shop drawings and other submissions
    Structural steel (detail drawings and erection drawings)
    Bar joists and prefabricated trusses
    Metal decking
    Reinforcing bar
    Product data
As-built drawings
Field and shop reports (including construction photos)
    Structural steel inspection laboratory (including weld and bolt inspection)
    Concrete inspection laboratory (reinforcing steel, formwork, concrete)
    Concrete mix designs
    Clerk of the works
    Structural engineer
    Architect
    Construction supervisor's daily log
    Local building inspector
    Owner's or developer's field inspectors
Materials Strength Reports or Certification
    Concrete compressive strength
    Masonry prism strength
    Steel mill certificates
    Welding procedures (e.g., type of electrodes, required preheat)
    Fastener certification
    Results of special load tests
Project correspondence[b]
    Owner/consultant
    Intraconsultant
    Owner/contractor
    Consultant/contractor
    Transmittal/records

Table 8.7 *(continued)*

In-house memoranda
Records of meeting notes
Records of telephone conversations
Consultant Reports
 Feasibility studies
 Progress reports
 Soils consultant reports (including boring logs)
Calculations
 Primary structural engineer
 Reviewing structural engineer
 Specific subcontractor's engineers (where required by contract)
Maintenance and modification records

 [a] Assist principally in establishing dead loads.
 [b] The scope will vary depending on the investigator's assignment.

opinions regarding procedural responsibilities for the failure, by indicating what information was available to those involved in the design and construction of the facility or what research and testing went into the development of a construction system or material. Common sources of such technical information include books, technical journals, technical and trade magazines, manufacturers' or systems suppliers' literature or research reports, proceedings of conferences and symposia, publications of trade organizations, university or research facility reports, and unpublished reports of individuals.

Computer literature search facilities accessible from remote terminals are becoming more common and useful. Two of the database services that the author has used with success are the Compendex System and the ASCE Civil Engineering database. Many large libraries now have computer search capabilities.

Table 8.8 Sources of Project Documents

Architects and engineers involved in original design, modification, or repair of facility
Past and present owners
Past and present tenants
General contractor and/or construction manager for original construction,
 modification, or repair of facility
Subcontractors involved in original construction, modification, or repair of facility
Developer of facility
Construction mortgagee of facility
Materials or systems suppliers for original construction, modification, or repair of
 facility
Previous or other current investigators
Building department
Testing agency involved in original construction, modification, or repair of facility

The Architecture and Engineering Performance Information Center, based at the University of Maryland, is a repository of performance information on constructed facilities. Project citations are accessible by computer terminal. Hard copies of documents are also available. Further information on AEPIC is given by Vannoy (1983).

8.7 THEORETICAL ANALYSES

The common purposes of performing structural and geotechnical analyses in investigations are to determine the causes of failure and to establish the degree of conformance of the structural and geotechnical design to applicable standards.

In design review, do not perform needless checks of components or regions of the construction that have nothing to do with the failure. Also, do not get hung up in exhaustive documentation of simple code violations that have no causal relationship with the failure (Janney 1986).

Structural analyses may attempt to determine stress, strain, strength, deflection, dynamic response (transient or harmonic), fatigue, fracture, or stability. Geotechnical analyses deal with strength, soil pressures, long-term and short-term settlements, and slope stability. Sophisticated soil–structure interaction analyses are now commonly performed.

For calculations to determine the causes of failure, analyses that include geometric and material nonlinear behavior frequently are necessary. Computer codes for finite-element and finite-difference analysis, such as NASTRAN, STRUDL, ANSYS, ADINA, and BOSOR, are readily available for these analyses. Secondary stresses from the effects of temperature, humidity, creep, shrinkage, foundation settlement, stresses induced during construction, and joint eccentricity are frequently ignored in structural design and often play a part in structural failures. Do not neglect these in your analyses. When the overall distribution of forces within the structure has been established and the problem is reduced to determining the resistance of a particular member, connection, or geotechnical element, three options are available:

1. Hand solutions (often based on previously performed research)
2. Finite-element or finite-difference analyses
3. Physical mockup testing

With simple hand solutions, scrutinize every parameter in the resistance relationship to see if the value you are using represents accurately the true as-built condition of the structure at the time of failure. For example, suppose the initiating cause of collapse of a concrete flat-plate structure was found to be punching shear, and the investigator had become convinced that the following ACI Code relationship accurately quantified that resistance:

$$V_c = 4 \sqrt{f'_c}\, b_o\, d$$

It first would be advisable to review the literature for the research on which this relationship is based. This might show that the factor of 4 more realistically should be taken to be 4.4. Next, determine the real f'_c at the time of failure. What was the real d? (Do not assume it was as indicated on the drawings.) Finally, examine the actual failure surface to get a better estimate of b_o.

Avoid blind overreliance on complex computer methods. Each finite-element analysis should be checked for satisfaction of overall equilibrium and should be scrutinized for qualitative response. Simple and approximate checks by hand solution to complex computer models should be made.

Probabilistic reliability analyses that account for the variation in the parameters the engineer has estimated for strength and resistance are becoming more common in failure investigations.

Where possible, establish structural resistances of key components or assemblies by both analytical and experimental methods. Correlation of the results of the two is persuasive proof of your findings.

8.8   LABORATORY TESTS

8.8.1   General

Books of standard tests of the ASTM and others for construction materials and assemblies fill shelves of libraries; other ad hoc tests that are never documented and disseminated are developed for specific projects. It is not the purpose of this section to describe these tests in exhaustive detail, but rather to generally describe some of the common types of laboratory tests and set forth some general considerations.

Based on the author's experience, the planning and conduct of laboratory tests is fraught with opportunities for error and invalid results. To avoid such pitfalls, follow these principles:

Understand the intent and purpose of the test you are conducting and its significance to your investigation before implementing it. The author has seen numerous investigators conduct standard tests just because they exist and were convenient to carry out, only to waste time and money, destroy evidence, produce misleading conclusions, and embarrass themselves under later scrutiny.

Whenever possible, witness the test yourself. If not possible, see that the test is conducted by a qualified professional who can testify as to the procedures used and the accuracy of results obtained.

Understand the relevant parameters that may cause results to vary from the in situ condition you are trying to represent. When variables cannot be established deterministically, consider parametric studies or sensitivity analyses to understand the effect of variables.

Choose the number of tested samples to be consistent with your objectives for level of confidence and type of determination desired.

Check calibration of all components.

Use redundant checks by alternate means of test methods.

8.8.2 Structural Laboratory Tests

Some commonly used structural laboratory tests are given in Table 8.9. Further general discussion is provided here.

Table 8.9 Common Laboratory Tests in Structural Investigations

Test	Reference (where applicable)
Component or mockup load tests	
Wood trusses	ASTM E1080
Wall, floor, and roof panels	ASTM E72
Shear resistance of framed walls	ASTM E564
Window/wall assemblies	ASTM E330
Data reporting	ASTM E575
Beam flexural strength	ASTM E529
(see also Table 8.5)	
Concrete materials	
Cylinder compressive strength	ASTM C873, C39
Modulus of elasticity	ASTM C469, C215
Thermal expansion	ASTM C531
Bond strength	ASTM C234
Tensile strength	ASTM C496
Flexural strength	ASTM C192, C42, C1018, C293, C78
Diagonal shear strength	
Fatigue strength	
Fracture characteristics	
Petrographic analysis	ASTM C295
Air content	ASTM C457, C138, C231, C173
Chemical analysis of cement	ASTM C114
Cement content	ASTM C85
Alkali reactivity	ASTM C289
Abrasion resistance	ASTM C779, C944, C418
Absorption	ASTM C642
Density	ASTM C1040
Metal materials	
Tensile tests	ASTM E8
Charpy impact	ASTM E23, A370, E812, A673
Hardness	
Compressive testing	ASTM E9
Ductility	ASTM E290
Acoustic emission	ASTM E1139

Table 8.9 (continued)

Test	Reference (where applicable)
Metallography	ASTM E807, E7, E112, E2, E883
Chemical tests	ASTM E60, A751
Corrosion	ASTM E937
Elongation	ASTM E8
Fatigue	ASTM E647, E812, E468, E467, E466, E1150
Masonry materials	
Compressive strength of units	ASTM E447, C67
Prism strength	ASTM C349, E447
Flexural strength	ASTM C1072, C348, C67
Bond strength	ASTM E518, C952
Shrinkage	ASTM C426
Mortar strength	
Shear strength	ASTM E519
Thermal expansion	ASTM C531
Tensile strength	ASTM C1006
Water absorption	ASTM C67
Efflorescence	ASTM C67
Freeze–thaw resistance	ASTM C67
Petrography	
Mortar air content	ASTM C1072
Chemical resistance	ASTM C279
Wood materials	
Compression strength	ASTM D2555, D143
Flexural strength	ASTM D1037, D198
Shear strength	ASTM D1037, D198
Tensile strength, modulus of rupture	ASTM D2555
Creep	
Shrinkage	
Moisture content	
Durability of adhesives	ASTM D3434
Weld inspection (see Table 8.5)	
Subsurface tests and nondestructive weld testing (see Table 8.5)	
Model tests	
Structural load tests	Schreiver (1980)
Boundary layer wind tunnel tests	
Water and air penetration	
Window/wall air leakage	ASTM E283
Window/wall water leakage	ASTM E331, E547, E1105, AAMA 501.3
Scanning electron microscopic examination	

8.8.2.a Component or Mockup Load Tests

These are similar to that described in Section 8.5 for site investigations. Tests may be performed on actual samples removed from the debris, from laboratory-built mockups, or from both (Figures 8.13 and 8.14).

8.8.2.b Concrete Materials

Tests include the mechanical properties of compression strength; modulus of elasticity; thermal expansion; abrasion resistance; bond, creep, and shrinkage characteristics; and tensile strength, flexural strength, shear strength, fatigue strength, and fracture properties (Whitehurst 1966; Malhotra 1976). Material tests include chemical analyses, petrographic analyses, and air content.

8.8.2.c Resistance to Environmental Attack

Various tests for this have been developed.

8.8.2.d Metal Materials

Tests of mechanical properties include yield strength, tensile strength, shear strength, creep, modulus of elasticity, ductility, elongation,

Figure 8.13 Full-scale laboratory load test of foam-filled sandwich panel.

Figure 8.14 Laboratory test for time-to-failure in acid environment.

fatigue properties, fracture toughness (Roberts and Pense 1980), and hardness. Metallography is used to confirm mechanical characteristics and environmental performance. Resistance to corrosion is determined by a number of methods. Chemical tests are also standardized.

8.8.2.e Masonry Materials

Masonry prisms, individual units, or mortar and grout materials are tested for mechanical properties of compressive, tensile, bond, and shear strength; modulus of elasticity; and volume changes caused by temperature, shrinkage, and humidity. Various petrographic and chemical analyses are performed to ascertain composition. Other tests include water absorption, freeze–thaw resistance, efflorescence, and resistance to other environmental effects.

8.8.2.f Wood Materials

Wood is an orthotropic material, and its mechanical properties are sensitive to duration of load. Mechanical properties tested include compression; shear, tensile, and flexural strength; modulus of elasticity; creep; and shrinkage. Microscopic and chemical analyses are performed for structure, composition, and resistance to decay. Dimensional stability under cyclic changes in moisture content is an important property.

8.8.2.g *Subsurface Tests and Nondestructive Weld Testing*

See Section 8.5.

8.8.2.h *Model Tests*

Structural model load tests, similar to component testing described earlier, are conducted using well-documented principles of similitude. Boundary layer wind tunnel testing of models is now commonly and readily employed to study wind pressures and suctions on building frames and cladding, flow directions, effects of wind on pedestrians, and aeroelasticity.

8.8.2.i *Water and Air Penetration, Heat Loss*

See Section 8.5.

8.8.2.j *Scanning Electron Microscope Examination*

SEM analyses are now commonly employed for material composition studies (Figure 8.15).

Figure 8.15 Scanning electron microscope view of nickel–sulfide inclusion, which caused fracture of tempered glass sheet.

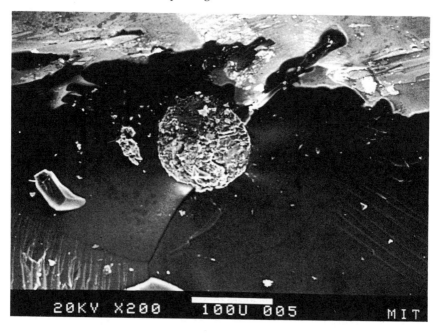

8.8.3   Geotechnical Laboratory Tests

Tests of soil and rock samples have become standardized through the ASTM and other systems. The most commonly employed methods are outlined here and in Table 8.10.

8.8.3.a   *Soil Classification*

Soil samples are classified under the Unified Soil Classification System (USCS) on the basis of their appearance, texture, and plasticity.

8.8.3.b   *Strength Tests*

These vary with the type of soil and property desired and include direct shear, triaxial compression, and unconfined compression tests.

Table 8.10   Common Laboratory Tests in Geotechnical Investigations

Test	Reference (where applicable)
Soil classification	
USCS classification	ASTM D2487
Strength tests	
Direct shear	
Triaxial shear	ASTM D3397
Triaxial compression	ASTM D2850
Unconfined compression	ASTM D2850
Water-related tests	
Consolidation	ASTM D4186
Moisture content	ASTM D2216
Permeability	ASTM D4525, D2434
Capillarity	ASTM D2325, D3152
Groundwater tests	
Hardness	
pH	ASTM G51
Sulfate or salt contact	
Others	
Specific gravity	ASTM D854
Grain size	ASTM D422, D1140
Compaction	ASTM D4523, D4254
Atterberg limits	ASTM D4318
Modulus and damping	ASTM E4015

8.8.3.c *Water-Related Tests*

These include consolidation tests to measure long-term volume change under load of water-bearing soil samples, determination of moisture content, permeability, and capillarity.

8.8.3.d *Groundwater Tests*

These tests are used to determine hardness, pH, sulfate or salt content, corrosion potential, or other chemical characteristics of groundwater.

8.8.3.e *Others*

Miscellaneous tests on soil samples include determination of specific gravity, grain size, compaction, and Atterberg limits.

8.9 FAILURE HYPOTHESES, DATA ANALYSES, FORMATION OF CONCLUSIONS

The activities described in this section are ongoing processes in the failure investigation. Failure hypotheses are developed early in the assignment and guide the investigation. However, the investigator must constantly be on guard not to take unconfirmed facts for granted, jump to conclusions, and close necessary avenues of investigation prematurely. As the investigation proceeds, some failure hypotheses may be eliminated; new hypotheses may be developed.

To establish the causes of a failure, the investigator must do two things:

1. Determine the mode and sequence of failure
2. Establish that for the initiating location of failure and for each successive step in sequence of failure the demands on the facility (e.g., loads, environmental factors, or input) exceeded its ultimate capacity (e.g., strength, stiffness, or durability)

For some cases, the first step may be trivial (e.g., "The concrete beam cracked diagonally three feet from its southern end" or "The settlement at Column D-2 exceeded one inch"). In cases of progressive collapse, determination of the initiating point, mode, or trigger can be difficult. Often, failure is initiated locally, but collapse occurs under general instability. Redundant structures are remarkable for their ability to overcome initial gross overstresses by shedding load through other paths. When gross failure occurs, the investigator must show that the resistance of every potential load path or failure mode has been exceeded by the load effects.

Civil engineering works fail for one or more of the following reasons:

Inadequate design

Improper construction

Inadequate materials or systems

Improper maintenance or operation

Failure of building codes or accepted engineering practice to recognize properly a certain demand or limit of capacity

Imposition of extreme or abnormal demands that the facility was not intended to resist

Experience shows that the last two infrequently cause failures. Notable exceptions are the failure of most building codes to recognize the effect of drifting snow until the early 1970s (Figure 8.16) and deficiencies in state-of-the-art practice in parking garage design to provide a durable facility. More often than not failures are caused by a number of interrelated factors.

With respect to strength issues, the true ultimate capacity must be exceeded for a failure to occur. It is not enough to show that the safe working load factored by the code factor of safety was exceeded. Structures often have substantial capacity beyond their code-defined ultimate capacities, and, as mentioned above, alternate load paths and forms of resistance, not recognized in conventional structural design, are often available.

Sometimes you will find that you will have trouble explaining

Figure 8.16 Roof collapse under drifting snow, shopping mall, Ohio.

not why the facility failed, but rather why it did not fail earlier. This is a clue that you have missed something in the load effects, the resistances, or the behavior. Often you find that there is an alternate load path or supplementary source of resistance that you missed.

The process by which failure hypotheses are developed and tested varies, of course, with project type. For complex investigations requiring several team members this may involve discussions at several steps of the investigative plan. Often there is an initial site investigation, failure hypotheses are set forth, data are collected, data are analyzed, failure hypotheses are revised, there is more site investigation, and so forth. Experience is invaluable in developing failure hypotheses, as facilities of a certain design or construction tend to fail in repeated modes. Many of these failure profiles are documented (Addleson 1982; Janney 1986).

In developing and refining hypotheses, do not narrow the focus of the investigation too soon; in particular, do not limit the investigation to likely or apparently obvious causes of failure. The early activities of each initial site investigation should be directed at establishing failure hypotheses. Keep all avenues open until they can be eliminated by positive proof.

At each stage encourage active debate of each of the issues among the team members.

Occasionally it may not be possible to pinpoint a single mode and cause of failure. Even the best investigators, for lack of information or other causes, are stumped sometimes. In such cases it is entirely legitimate, and indeed it is incumbent on the ethical investigator, to present opinions in terms of a number of most likely modes and causes of failures, along with an opinion of their relative likelihoods.

8.10   DETERMINATION OF PROCEDURAL RESPONSIBILITIES

When an investigative case is involved with litigation or other forms of dispute resolution, the forensic engineer is usually called on to provide opinions on the responsibilities of various parties (defendants) for the failure. Such opinions, of course, are based on the investigator's judgment and expertise as a professional engineer; the engineer should not and cannot draw legal conclusions. These judgments can be extremely difficult to make, requiring the utmost integrity, objectivity, and impartiality. In the author's experience, disagreements between experts in dispute resolutions are more frequently over opinions of responsibility than the technical causes of failure. Some recurring procedural issues are discussed in ASCE (1973), Stockbridge (1973), Thornton (1982, 1983), Gibble (1986), Ellingwood (1987), and Sowers (1987).

The engineer is normally required to provide opinions based on a professional "standard of care," which may be a different standard than that of his or her own firm. Definitions of standard of care vary with legal jurisdiction, but generally revolve around the concept of "the actions of a reasonable and prudent professional acting under like or similar circumstances, practicing at the same approximate time and geographic location." The nuances of varying definitions can be important, however. Check with your attorney for each case.

Conformance of original design and construction to applicable codes, regulations, and other written standards must be determined on the basis of those standards that applied to the particular project at the time of original construction (usually the time of issuance of the building permit). The investigator should carefully research which standards are applicable before offering such opinions; the author has seen many investigators err on this point. Simple lack of compliance to codes or standards does not necessarily prove negligence. The expert must show a causal connection between the lack of compliance and the actual damage.

Frequently, there are multiple procedural causes of failure; no single firm or individual may be responsible for the damage. In such cases, determination of relative responsibilities is difficult. When responsibility may be mixed, the investigator should check with the attorney to see what questions of allocation of responsibility need to be addressed. Is it sufficient to identify a single action of negligence that is "substantially responsible" for the failure, if such a cause exists? Or, are all contributory causes relevant, no matter how minor the contribution? A useful analysis tool in such cases of mixed responsibility is to consider "what-if" scenarios, wherein it is hypothesized that a certain component of negligence did not exist: "What if the design was perfect, but all construction defects existed?" "What would the result have been?" "Suppose the construction was perfect and only the identified design deficiencies existed?"

8.11 REPORTS

The report is the culmination of all previous efforts. It is the final (and possibly only) product the client sees; and, if the case is part of a lawsuit, it is the basis of the expert's testimony. Poorly written, it can be a source of unending difficulty for the expert witness. In addition to generating the end product of the investigation, report writing can be a valuable step in the investigative plan because it requires that information from the several team members be brought together, and it forces the writer to organize and review all issues.

The requirements for good report writing for structural investigations are fundamental writing skills of grammar, syntax, style, punctuation, and usuage, and organization. General fundamentals of technical writing skills are covered in Chapter 11. Painstaking effort should be directed to the report. Table 8.11 shows a general outline for report organization that has withstood the test of many years of varied projects at the author's firm. Several of these topics are discussed here.

Table 8.11 General Report Outline

Letter of Transmittal
Abstract
Table of Contents
1. INTRODUCTION
    1.1 Objective
    1.2 Scope
    1.3 Background
    1.4 Responsible Design and Construction Agencies
    1.5 Construction Documents
2. DESCRIPTION OF STRUCTURE (OR PROJECT)
3. FIELD INVESTIGATION
4. LABORATORY TESTS
5. RESULTS OF CALCULATIONS
6. DISCUSSION OF FIELD INVESTIGATION, LABORATORY TESTS, AND RESULTS
    OF CALCULATIONS
7. CONCLUSIONS
8. RECOMMENDATIONS

Abstract. Limit this to one page, except in rare complex cases. Describe as concisely as possible what the report covers and give a summary of principal conclusions.

Objective and Scope. Often omitted from reports, these sections are essential to set forth why the investigation was undertaken and what work was performed.

Background. The history of the development of the project is presented here. All information obtained from others that has not been substantiated by the investigators is also included. If you must use information obtained from others in developing opinions and conclusions, say so.

Responsible Design and Construction Agencies. This is simply a listing for background.

Construction Documents. This, also, is a listing, not a description of contents.

Field Investigation. Include names of investigators and dates of visits. Write in the first person: "I saw . . . " not "It was observed . . . ." This section should include only facts and observations, not opinions.

Laboratory Tests. As with field investigation, only facts should be reported in this section. Interpretation of test results is included in the discussion (next section of the report).

Discussion of Field Investigation, Laboratory Tests, and Results of Calculations. This is the first section that should contain interpretation of findings and opinions. All discussion must be based on factual information presented in previous sections of the report. In long reports, this section should be subdivided.

Conclusions. The conclusions should flow logically from, and be supported entirely by, the discussion. Use a concise numbered or bulleted list of items.

Recommendations. This section is not appropriate for all investigations. When given, recommendations should be based only on other matter in the report.

The entire report must present a convincing and logical argument from facts to discussion to conclusions. The casual relationship between identified deficiencies in design, materials, or construction and the actual failure must be shown.

On small or simple projects or for progress reports, a shorter, less formal report, or letter report may be appropriate. The letter report follows the same organization as described here, but the letter of transmittal and abstract are excluded.

References

Addleson, L. 1982. *Building Failures—A Guide to Diagnosis, Remedy and Prevention.* London: Architectural Press.

American Welding Society (AWS). 1980. *Welding Inspection.* Miami, FL: AWS.

_____ 1986. *Guide for Nondestructive Inspection of Welds*, ANSI/AWS B1.10-86, Miami, FL: AWS.

American Society of Civil Engineers (ASCE), Research Council on Performance of Structures. 1973. *Structural Failures: Modes, Causes, Responsibilities.* New York: ASCE.

_____ 1988. *ASCE Official Register 1988.* New York: ASCE.

Ang, A. H. S., and W. H. Tang. 1975. *Probability Concepts in Engineering Planning and Design.* New York: John Wiley and Sons.

Bell, G. R., and J. C. Parker. 1987. Roof collapse, Magic Mart Store, Bolivar, Tennessee. *Journal of Performance of Constructed Facilities* (ASCE) 1, No. 2 (May).

Bendat, J. S., and A. G. Piersol. 1971. *Random Data: Analysis and Measurement Proceedings.* New York: John Wiley and Sons.

Betz, C.E. 1986. *Principles of Penetrants.* Chicago: Magnaflux Corp.

Blockley, D. I. 1981. Logical analysis of structural failure. *Journal of the Engineering Mechanics Division* (ASCE) 107, No. EM2 (April).

Colloms, M. 1983. *Computer Controlled Testing and Instrumentation.* New York: John Wiley and Sons.

Crist, R. and J. Stockbridge. 1985. Investigating building failures. *Consulting Engineer* (New York) (June).

Ellingwood, B. 1987. *Design and construction error effects on structural reliability, Journal of Structural Engineering* (ASCE) 113, No. 2 (February).

Fairweather, V. 1975. Bailey's Crossroads: A/E liability test. *Civil Engineering* (ASCE) 45, No. 11 (November).

Feld, J. 1968. *Construction Failure.* New York: John Wiley and Sons.

Gibble, K. ed. 1986. *Management Lessons from Engineering Failures.* New York: ASCE, October.

Janney, J. R. 1986. *Guide to Investigation of Structural Failures.* New York: ASCE.

Jones, L., and A. F. Chin. 1983. *Electronic Instruments and Measurements*. New York: John Wiley and Sons.

Kaminetzky, D. 1976. Structural failures and how to prevent them. *Civil Engineering*. (ASCE) 46, No. 8. (August).

Kaminetzky, D. 1983. Failure investigations, who? why? when? what? and how? *Proceedings, Construction Failures: Legal and Engineering Perspectives*, American Bar Association, October.

LePatner, B. B., and S. M. Johnson 1982. *Structural and Foundation Failures: A Casebook for Architects, Engineers, and Lawyers*. New York: McGraw–Hill.

Loomis, R. S., R. H. Loomis, R. W. Loomis, and R. W. Loomis. 1980. Torsional buckling study of Hartford Coliseum. *Journal of the Structural Division* (Proceedings of ASCE, Vol. 106, No. STI, Proc. Paper 15124, January).

Loss, J., and E. Kennett. 1987. *Identification of Performance Failures in Large Scale Structures and Buildings*, NSF Project No. ECE 8608145. College Park: University of Maryland.

Malhotra, V.M. 1976. *Testing Hardened Concrete: Nondestructive Methods*. Detroit: American Concrete Institute.

National Academy of Sciences (NAS), National Academy of Engineers. 1969. *Full-Scale Testing of New York World's Fair Structure*, Vol. I–III. Washington, DC: NAS.

National Bureau of Standards (NBS). 1982a. *Investigation of the Kansas City Hyatt Regency Walkways Collapse*, NBS-BSS 143. Washington, DC: U.S. Department of Commerce, May.

_____ 1982b. *Investigation of Construction Failure of Harbour Cay Condominium in Cocoa Beach, Florida*, NBS-BSS 145. Washington, DC: U.S. Department of Commerce, August.

_____ 1982c. *Investigation of Construction Failure of Reinforced Concrete Cooling Tower at Willow Island, WV*, NBS-BSS 148. Washington, DC: U.S. Department of Commerce, September.

_____ 1987. *Investigation of L'Ambiance Plaza Building Collapse in Bridgeport, Connecticut*, NBSIR 87-3640. Washington, DC: U.S. Department of Commerce, September.

Pfrang, E. O., and R. Marshall. 1982. Collapse of the Kansas City Hyatt Regency Walkways. *Civil Engineering* (ASCE) 52, No. 7 (July).

Roberts, R. and A. Pense. 1980. Basics in failure analysis of large structures. *Civil Engineering* (ASCE) 50, No. 5 (May).

Ropke, J. C. 1982. *Concrete Problems: Causes and Cures*. New York: McGraw–Hill.

Ross, S. S. 1984. *Construction Disasters: Design Failures, Causes, and Prevention*. New York: McGraw–Hill.

Schousboe, I. 1976. Bailey's Crossroads collapse reviewed. *Journal of Construction Division* (Proceedings of ASCE, Vol. 102, No. C02, Proc. Paper 12186, June).

Schreiver, W. R. 1980. *Full-Scale Load Testing of Structures*, ASTM STP 702. Philadelphia: American Society for Testing and Materials.

Smith, D. L. 1985. *Test and Measurement Techniques Using Personal Computers*. Madison, NJ: American Institute for Professional Education.

Smith, E. A., and H. I. Epstein. 1980. Hartford Coliseum roof collapse: Structural collapse sequence and lessons learned. *Civil Engineering* (ASCE) 50, No. 4 (April).

Sowers, G. F. 1987. Investigating failure. *Civil Engineering* (ASCE) 57, No. 5 (May).

Stockbridge, J. G. 1973. Cladding failures—Lack of a professional interface. *Journal of the Technical Councils* (Proceedings of ASCE, Vol. 43, No. 1, Proc. Paper 15085, January).

Thornton, C. H. 1982. Lessons learned from recent long span roof failures. *Notes from ACI Seminar on Lessons from Failures of Concrete Buildings*, Boston: American Concrete Institute, April.

_____ 1983. Catastrophic failures—Why they happen. *Proceedings—Construction Failures: Legal and Engineering Perspectives*, American Bar Association, October.

Vannoy, D. W. 1983. "20/20 hindsight—Overview of failures. *Proceedings—Construction Failures: Legal and Engineering Perspectives*, American Bar Association, October.

Whitehurst, E.A. 1966. *Evaluation of Concrete Properties from Sonic Tests*. Detroit: American Concrete Institute.

9. Environmental Systems Failures

FRED H. TAYLOR, P.E.

9.1 INTRODUCTION

The subject of "environmental systems" normally refers to buildings and the systems that provide a comfortable and stable environment for their occupants, that is, air conditioning and lighting systems. However, the field is much broader. It is broken down into three general areas:

> Building environmental systems, that is, HVAC, lighting, acoustics, and so forth
>
> Industrial and process environmental conditioning
>
> Microclimate conditioning for specific processes

In each of these areas, the environment is supported by a myriad of individual components that make up the conditioning system, any one of which can precipitate a complete system failure.

The opportunity for failure in these systems is directly proportional to the number of individual components and subcomponents. The probability of failure is a function of the reliability factors that go into the system, from concept to the end of the system's life. The consequences of failure range from mild discomfort to severe economic loss, from illness to injury or death.

The challenges to an environmental systems forensic engineer (ESFE) are immense. The forensic engineer is often looking at the failure of an end product. Therefore, the engineer must be capable of tracing the way back from the failure point through its operating history, construction, design, all the way to its conception. The root cause of the failure may be found any place along the way and is, more often than not, camouflaged by a number of suspicious but unrelated symptoms. This requires a full command of the physics, processes, systems, details, and requirements of the system and related disciplines. In addition, there are those alleged or actual failures that occur from human misuse.

The point where the environmental system failure occurs establishes the order of events leading to the services of the forensic engineer. The impact of the failure determines the course of the investigation and its final disposition. If the system failure results in an overheated office and all of the occupants have to go home early (and this is a repetitive problem), the forensic engineer's job may simply be to identify the cause of the problem, report to the client, and make recommendations for corrections or revisions so that the problem will not reoccur. The work may end at this point.

If the failure is a boiler explosion resulting in loss of life and great physical and economic damage, the forensic engineer, and all of his or her work, may well end up in court. In that case, the path to the failure and beyond must be traced with faultless logic and the work must be able to stand up against relentless examination.

The interesting dilemma in forensic work is that the seemingly simple office overheating problem may escalate into a major confrontation between the affected parties, where the forensic engineer's work will have to withstand the same relentless examination as in the case of the catastrophic failure. The investigation may not be as extensive but it must be as accurate.

The type of environmental systems failure also determines the type of client. In the case of the overheated office, the client may be the building owner, who is intent on reoccupying the building and finding and correcting the cause. Or the client may be the workers' labor organization, intent upon pursuing a grievance. In the case of the catastrophic boiler explosion, the client may be the attorney for the insurance company, the boiler or combustion control manufacturer, or the building owner.

The various client–engineer relationships are discussed in greater detail later in this chapter.

9.2 PURPOSE AND SCOPE OF INVESTIGATIONS

9.2.1 Functions of the ESFE

An ESFE is a diagnostician, detective, reporter and "expert." The ESFE may also find him/herself in the role of corrective design engineer. If this sounds like nothing more than system troubleshooting and failure analysis, that is to be expected. The work becomes forensic engineering when the system failure causes someone harm and has the potential to end in litigation, where the investigator becomes an expert witness.

The only consistent elements between various projects are that there has been a failure and that the path to finding it will be a new and challenging experience, every time.

My experience with forensic consulting is as a professional consulting mechanical engineer whose primary field of expertise is in facilities. Consequently, those areas where I am most qualified and conversant are HVAC systems, temperature controls, industrial ventilation, and processes embodying these disciplines. Therefore, these fields are discussed in this chapter to explain the functions of the ESFE. The principles are applicable to all other disciplines in the environmental field such as lighting, power, acoustics, and waste disposal.

To demonstrate the functions of the ESFE, a case study is used. The project involved an air conditioning system failure in a 44-story high-rise office building in San Francisco, California.

This building which had been completed and occupied for about one year, had chronic overheating problems during the cooling season that were not resolved by the mechanical contractor or mechanical design engineer. The architect, facing an unhappy client as well as a potentially serious liability problem, retained me to investigate the HVAC system and find the cause of overheating.

This project demonstrates the role of the ESFE, relationships with the various parties, the path and techniques followed in the investigation, and the many false trails that present themselves.

As an ESFE, I conducted a detailed investigation of the building over a period of several months. At the onset, a series of preconceived convictions by the building owner, the building operating engineers, the mechanical design engineer, and the many self-styled experts occupying the building were investigated. These convictions held, among other things, that the zone temperature control induction boxes supplying conditioned air to the space were faulty and would not shut off heat to the space. There was also concern that the air conditioning system design was inadequate and not capable of providing sufficient cooling.

The owner, concentrating on the symptoms, was preparing to issue a $250,000 (in 1976 dollars) contract to make an expedient correction to resolve the complaints. (This would not have corrected the root cause of the problem and would have identified a liability amount for a claim against the architect.)

The final outcome of the investigation was the discovery of an obscure crossover pipe between the building heating hot water system and the high-temperature domestic hot water heating system. This crossover caused hot water to be inducted into the building heating system from the domestic hot water heating system, even when the building heating pump was shut down and the supply flow shut off.

Along the way to this discovery, a whole series of related conditions and problems were discovered. Some contributed to the problem whereas others proved to be false leads that had no impact on the problem.

To accomplish the investigation, the following steps were taken:

1. The building was inspected to get the "lay of the land" and to become familiar with the various systems.
2. The building design plans and other contract documents were studied to determine the construction requirements and intent.
3. Cursory load calculations were made to confirm whether the basic design concepts were reasonable.
4. A series of tests was formulated and carried out on the various systems and subsystems. These tests were conducted as if they were academic laboratory experiments; rules of evidence, logic, and closure were used.
5. Regular meetings were held with the architect, design engineer, and owner to keep them apprised of the progress of the investigation. Great care was taken so as not to state conclusions unless they had been proven; even then, all conclusions were labeled as preliminary until the final report was prepared. Because the affected parties were kept informed, full support was maintained. All parties knew that all of the evidence was available to all parties and that nothing was being concealed for anyone's protection or benefit.
6. At the conclusion of the investigation, a full written report was prepared, delineating the study findings. These findings not only revealed the root cause of the problem and a solution; they also listed all of the other "conditions" discovered. These were of great benefit to the owner.

With the discovery and correction of the root cause of the problem, the specter of litigation disappeared, the building owner and occupants were satisfied, and the owner requested the ESFE's firm to provide continuing facility engineering services. In other words, everyone came out a winner. This case study is discussed in further detail in Section 9.7.1.

9.2.2 Client Relationships

There are two common entrances to the client relationship: (1) retention directly by an affected client and (2) retention by an attorney looking for an expert witness.

The initial client in a environmental systems failure case is normally the one who is affected by the failure to the point where something must be done, and all normal channels have failed. From there, a number of engineer–client relationships can develop.

In the case of the 44-story office building, the owner was the injured party. The owner precipitated the investigation by demanding that the architect, the party responsible for the design and construction of the building, solve the problem. The contractor and his subcontractors, as well as the design mechanical engineer, had not been successful in their efforts. Therefore, the architect had to obtain an outside expert.

At this point, litigation was a distinct possibility. If the problem could not readily be resolved and was discovered to be a costly design or construction error, the outside expert would also have to be a potential expert witness, either in arbitration or in a court of law.

Once the services of the outside expert were secured, the first client–ESFE relationship was formed: in this case, the ESFE was the author—a consulting mechanical engineer qualified in HVAC systems.

After completion of the investigation and report, a second client–engineer relationship formed: my consulting engineering firm entered into a facility engineering relationship with the building owner. This relationship continues to the present time.

At the conclusion of the investigation, the person who knew the most about the building, its systems, problems, and operations, was the ESFE, in this case, myself. I was the logical person to continue with the required facility operating and maintenance engineering.

The foregoing discussion leads to the subject of *Ethics*. The environmental system analysis often identifies a problem to be corrected. If the ESFE also practices engineering design, he or she may be the most qualified and logical person to perform the engineering required for the remedial work. This raises the ugly specter of *Conflict of interest*.

A definite conflict arises if the ESFE views each case as a source of continuing design business. Judgment will be biased. And yet, the ESFE, with preparatory knowledge beyond the scope of the report that should not be wasted, can be a valuable resource for the owner. Therefore, the ESFE faces a dilemma, a dilemma that can be resolved only through professional integrity.

Each client relationship must be compartmentalized. The original investigation must be conducted with eyes on that task only. Any corrective action recommended from a finding must be reported so that any other qualified engineer could be chosen to implement it. In the case of potential disputes, in no circumstances should it be implied that the ESFE is available for continued work.

Continued engineering can be ethically conducted only if initially solicited by the other parties after the ultimate disposition of the case, if no conflict of interest exists, and then only if agreed to by all parties. Integrity and ethics should control all professional relationships.

Many other client–ESFE relationships are possible depending on the situation. These can be with

Lawyers	Contractors	Other engineers	Plant engineers
Owners	Property managers	Architects	Corporations

Irrespective of the arrangement, all of these relationships must be conducted with integrity.

9.2.3   Fields of Practice

The fields of practice for the ESFE have been identified as the HVAC environment, industrial and process environmental conditioning, and microclimate conditioning for special processes. These are very broad. They are based on a pyramid of subsystems and equipment, each of which can be a field of practice within itself. The aspiring ESFE must have a command of each: either a close familiarity or a source of collaborating engineering.

The subfields of knowledge for the HVAC field, for example, are included in the *ASHRAE Handbook* index. This lists those topics for which the ESFE must have a command, a source of reference and/or support. This index comprises 36 $8\frac{1}{2}$ × 11 pages and lists 722 primary topics, with thousands of subtopics. The HVAC field alone is immense.

In addition to technical aspects of the subject, which are covered by the ASHRAE guides, there are building codes, state and local codes, and fire and life safety codes. These also have direct bearing on the work of the ESFE.

All of this knowledge must be combined with a familiarity of products, construction, industry standards and practices, and applications. Nearly any one of these topics can occupy a lifetime of study if a person wishes to specialize.

As such a wide range of knowledge and practice is required, the ESFE can never have a full command of the subject. He or she can only hope to have a general understanding of the scope of the field, as broad an area of expertise as possible, and the ability to research the subject and to find other experts outside his or her specialized expertise on the subject.

When the range of knowledge is extended beyond HVAC to electrical, chemical, and other environmental disciplines, it becomes apparent how extensive the field of practice is for the environmental systems forensic engineer.

Humans have developed environmental requirements or expectations for a person or a building's contents in almost any situation—in living and working quarters, in the food chain, in manufacturing processes, underground, in the sky, in and under the seas, and now in outer space.

9.2.4   Qualifications of the ESFE

The potential field of practice for the ESFE is significantly broad. The aspiring forensic engineer should realize that it is not possible to be an expert in all things and should know where to go to obtain qualified professional assistance.

As a practical matter, the HVAC engineer must first have a sound foundation in engineering basics, with a command of physics, hydraulics, psychrometrics, thermodynamics, fluid mechanics, and codes, backed up by credentials and experience. Above all, the successful forensic engineer must have experience.

Next are the analytical skills: the ability to observe a large mass of impressions, realize their individual and collective significance, evaluate them in the light of physics and psychology, and arrive at a rational conclusion. The engineer has to understand scientific analysis and rules of evidence. Every observation, measurement, calculation, and conclusion may be challenged at some point. Therefore, the first challenge should come from the investigator.

The most important tool that the ESFE must take into any investigation of a systems failure is an open mind. Remember, the goal is the truth. When the ESFE is retained for an investigation, typically the failure or its symptoms have existed for some time. Many people have been affected and they all have an opinion. This applies more often in environmental systems failure investigations than in other forensic work. Some of the occupants will have an ax to grind. Consequently, the ESFE receives lots of help, opinions, and directions, much of which will be of no value or may even be misleading. Seeds of golden truth are normally embedded somewhere in all that information. However, the truth cannot surface if buried by an overabundance of trivia, misinformation, or the investigator's preconceived ideas and prejudices.

The second most important tool is a sense of order about the investigation. Indeed, one of the most difficult aspects is the maintenance of order throughout the investigation in the midst of all the conflicting evidence, misleading symptoms, and other distractions. Often, several paths of investigation must be followed to dead ends before the truth is within reach.

The third item is the technical skills and background, supported by lots of experience, that are necessary to analyze the environmental problem at hand.

Fourth, but of equal importance, is the ability to write a clear report and to teach the client the technical aspects of the report so it can be comprehended. It all comes down to clear communication.

One of the greatest distractions in any investigation is the discovery of other systems failures that have no influence, or minor influence, on the failure under investigation. Because they are made of so many components, environmental systems can be expected to have one or more other failures unrelated to the primary failure. These other failures can wreak havoc in an otherwise orderly investigation.

Each anomaly discovered in the system must be entered and

evaluated to establish its place in the investigation. If not handled properly, a side issue may become a major issue.

9.3 TECHNIQUES

Good investigative technique is a key item in environmental systems failure analysis. At this point, the engineer is a detective. Guidelines for a typical investigation consist of the following:

1. *The complaint, where the client explains the problem.* The client may be a layperson and may not know what the problem really is. The client may see only the symptoms and may have strong misconceptions regarding the failure cause and needed remedial work. The ESFE must first identify the task and reach an understanding with the client. Then the problem can be pursued.

2. *A preliminary survey of the site, the mechanical system, and the facility.* The ESFE should walk around the building or site, especially the mechanical rooms. He or she should act like a sponge, absorbing information and using all of the senses.

3. *An interview with the affected parties.*

   Client—Refine the symptoms. Find out what the problem really is and how the client is affected. Identify the question.

   Occupants—They can be a valuable source of information. Talk to them and listen. The occupants often have information that the client does not know.

   Maintenance personnel—They have had experience with the problem, and often know the solution but are afraid to express it because of their insecurity or lack of technical expertise.

   Equipment suppliers—These individuals constitute a technical resource. Determine if the supplier agrees with the application or installation and if similar incidents have occurred before.

4. *Review of original plans and other contract documents.* Identify the designer's original plan and intent. This is very important. The designer's intent may not agree with the operator's understanding or requirements or with the owner's expectations.

5. *Detailed survey of site and system.*

   Investigate installation configuration.

   Investigate controls.

   Measure operating parameters.

Set up tests.

Look in all nooks and crannies.

6. *Engineering calculations.*

   Identify *actual* load.

   Calculate system capabilities.

   Quantify failure.

7. *Review equipment rating catalogs.* Is the equipment capable? Is it operating within its rating envelope?

8. *Analysis.* Review all input data.

   Back-check on survey if required.

   Back-check on interviews.

   Identify failure and circumstances.

   Address other related factors.

   Review entire investigative procedure to verify that this conclusion is more probable than any alternative theory.

   Be your own critic; challenge all of your conclusions.

   Assign responsibility, if applicable.

9. *Report.* In my practice, the report typically has three parts:

   Executive summary—Stating the problem and conclusions, written for a layperson.

   Technical section—Stating the findings, procedures, and logic to the conclusions.

   Technical support—To back up the findings.

   The report must be understandable to a broad spectrum of readers. Many readers will be laypersons who have little technical understanding of the conditions. At the other end of the spectrum are technical engineers as qualified as the ESFE, who will challenge the work.

   At this point, the important work is complete and the report becomes the authority. All future work should refer to this report.

10. *Delivery of report.* Explanation of technical report to laypersons.

    Be prepared to support and defend findings.

    Be prepared to teach, so that the client and others who do not have technical knowledge can understand the principles involved.

    Be prepared to proceed further if the client requires. In the case of nonlitigation projects, other services may be required, for example, recommendations for correction, preparation of plans, coordination of corrective implementation, tests, and system checkouts. All of these services can be provided within the framework of the highest level of ethics.

In the case of litigation, other required services might include support in litigation hearings, dispositions, and courtroom proceedings.

9.4 TOOLS

9.4.1 Tools in the Field

Environmental systems are normally supported by heating, cooling, ventilation, air flow, water flow, power, and controls. To quantify these phenomena, the following parameters must be measured:

Temperature
Mass flow (air and water)
Pressure
Equipment speed
Electrical voltage/amperage
Vibration/sound
Time

In investigations of this type, accuracy is of paramount importance. The calculations and resulting conclusions are only as good as the measurements on which they are based. There is nothing more disappointing than for a presentation to be shot down by a faulty measurement value. The credibility of the ESFE and of the entire case can be immediately destroyed; the most enlightened conclusion can be completely discredited. And the "egg" on the ESFE's face can be permanent.

To avoid this situation four precautions should be taken:

1. Be exact with measurements and conduct them in an acceptable manner. Use an acceptable published technique such as those accepted by the Associated Air Balance Council (AABC) and the American Society of Heating, Refrigeration and Air Conditioning Engineers (ASHRAE).
2. Use good instruments that are accurate and simple. They must retain their calibration throughout the investigation and later, if prolonged litigation is involved. All measurements should be repeatable.
3. Be able to trace the calibration of every instrument to a calibrated standard. Start each investigation with instrument calibration, and complete it with a back check to that calibration. Maintain a written record of these checks.
4. Set an accuracy range. When presenting values obtained by field measurement, never claim that they are absolutely accurate. It is difficult enough to maintain accuracy in a testing laboratory where the parameters are controllable. In the field, laboratory standards cannot be maintained and things just do

not work the same. Express the range of accuracy in all reports and in testimony.

Those instruments that have the closest relationship to basic principles of physics are preferred. The instruments I use most often are

Thermometers. A set of 16-inch glass/mercury thermometers accurate to within 0.2°F. These are useful for direct field measurements and also as a standard against which other specialty instruments can be calibrated. If a clean sock is placed over the bulb, accurate wet-bulb temperatures can be taken and, therefore, an accurate relative humidity can be obtained.

Pressure. 0- to 6-inch wc range—A set of three manometers using a standardized oil of known specific gravity. The first is a portable electronic point gauge-type manometer using a micrometer for readings as accurate as ±0.00025 inch wc from 0 to 2 inches wc. 0- to 16-inch wc Range—Dual-range U-tube incline manometer. These pressure instruments are supported by a collection of Pitot tubes for various duct sizes and hard-edge orifices.

Use of the preceding instruments is based on physical phenomena. These instruments are as accurate as is reasonable in the field, and are the least questionable. They all maintain their calibration and can usually be used as a standard for calibration of more sophisticated instruments. The instruments in the following group are still basic but require calibration, which can be done in the field.

Pressure (15 psi and above). Test-quality Bordon tube pressure gauges with recalibration adjustments.

Differential pressure. Diaphragm or Bordon tube gauges with bleed and equalizing values and connectors to the common test devices.

Air flow. Rotating vane anemometer and ceiling diffuser hoods.

Tachometer. Timed counter type.

The following instruments have manmade accuracy. As such, they require periodic factory calibration to ensure their continued accuracy. All must be of test quality so they can stand up to scrutiny.

Electronic thermometers
Electrical voltage: volt ohmmeter
Electrical amperage: clamp-on ammeter
Electronic digital thermometer
Recording thermometers
Recording electrical instruments

9.4.2   Tools in the Office

The excellence of the field tools must be matched by the intellectual and calculation techniques. The engineer must be well prepared—have a firm grounding in physics and a command of thermodynamics, mass flow, psychrometrics, mathematics, and English composition. The most brilliant findings or conclusions are worthless unless they can be communicated.

The most qualified and accepted references for environmental systems are the manuals published by the American Society of Heating, Refrigeration, Air Conditioning Engineers. These five manuals cover Fundamentals, Systems, Equipment, Applications, and Refrigeration and are commonly called the *ASHRAE Handbooks*.

ASHRAE is an international society of engineers dedicated to the advancement of the science of heating, refrigeration and air conditioning engineering, and related sciences. As such, it has developed data and is the leading authority on matters affecting building environments.

The *ASHRAE Handbooks* are supported by the annual *Transactions* containing the technical papers published by the society. These papers do not carry the weight of the handbooks but provide a conduit for the dissemination of individual research results to the industry. The evaluation of a systems failure may well require a search of the *Transactions* for relevant sources.

ASHRAE also publishes documents covering the work of the individual ASHRAE technical committees. Typical publications are the *Standards*, which describe uniform methods of testing, specify design requirements, and recommend standard practices. An excellent example is *ASHRAE Standard 61-1981: Ventilation for Acceptable Indoor Air Quality*. This standard recommends ventilation rates that provide healthful and comfortable indoor environments. It lists recommended outdoor air requirements for various building occupancies and uses.

Another valuable technical reference is *Industrial Ventilation— A Manual of Recommended Practice* published by the Committee on Industrial Ventilation of the American Conference of Governmental Industrial Hygienists. This manual is a guide to the practice of industrial ventilation wherein acceptable techniques are delineated. Also included is a list of *threshold limit values* for airborne concentrations of various substances and representable conditions under which it is believed that workers may be repeatedly exposed, daily, without adverse effect. Other useful information and values on physical materials and contaminants is also included.

Also considered to be qualified reference manuals are the design manuals published by the primary air conditioning manufacturers such as the Carrier Corporation and Trane Corporation. Although published by proprietary manufacturers, they are produced in a professional manner and are based on physics and engineering principles. They are not sales manuals and normally are not biased to the manufacturers' products. The manufacturers publish them to further the industry and, therefore, their own business opportunities. These manuals often become standard design references for field and office work, and represent an accepted standard of practice.

Other manufacturers of mechanical and electrical equipment publish catalogs covering their equipment. These usually contain the engineering ratings of the company's products along with the salient features and specifications. Technical products may also have extended rating publications, installation and operating manuals, and parts lists. These manufacturers' publications constitute a valuable source of reference material. Some of the most valuable may be the old catalogs of obsolete equipment. An ESFE should never let a manufacturer's representative remove one of these old catalogs from the office. Also, an effort should be made to keep a library of catalogs of equipment that the ESFE does not specify regularly in his or her own design work.

9.5 ASSOCIATED DISCIPLINES

The practice of environmental systems engineering is divided into three general areas:

1. *The system.* The total of all subsystems and components, how they work together, and how they accomplish their task.
2. *The system capacities.* The magnitude of these subsystems and the capacities of the components, all functioning together to provide a value.
3. *The application.* How the system is applied to solve a particular problem.

These three general areas are supported by the many engineering disciplines required by the individual systems, components, and applications, for example, thermodynamics, hydraulics, structures, architecture, and electricity. An environmental system encompasses too broad a field for any one engineer or discipline. It is, by its very nature, multidisciplinary.

In the HVAC field, the mechanical engineer provides the foundation. It is the mechanical engineer's skills and knowledge that support the creation of our environmental systems. As such, mechanical engineering is the lead discipline. However, the environmental system cannot work without electrical power, both for the prime movers and for the controls. Consequently, the electrical engineer is the closest associate in both the design of a new system and the investigation of a failure. Also, the electrical engineer is normally responsible for lighting, which is an environmental issue in its own right.

The association between mechanical and electrical engineers is interesting. The mechanical engineer must have a good foundation in electrical power and understand almost as much about the subject as the electrical engineer, insofar as it relates to the operation of the mechanical equipment. But the mechanical engineer is not the authority by disipline definition. The electrical engineer's expanded skills and knowledge are an indispensable support for the mechanical engineer.

As an example, a few years ago my firm was requested to find the cause of a fluorescent light flicker in a classroom. The flicker occurred only when the adjacent air conditioning compressor was operating and this compressor was immediately suspected.

The school air conditioning system consisted of a 25-ton package chiller/evaporative condenser combination unit that provided chilled water to fan coil units in the classroom. The refrigeration compressor in the package chiller was an accessible hermetic type, where the electric motor was directly connected to the compressor within a cast iron housing. The cold refrigerant suction gases passed over the motor on the way to the compressor crankcase and pistons, thereby cooling the motor. Because of this cooling, the motor could be smaller than an external air-cooled motor.

The motor had another unique feature: it had a much smaller diameter armature than units built by competitive manufacturers that did not suffer from similar flickering.

An implied relationship existed between the lights and the air conditioning equipment, but measurements with normal field instruments gave no confirmation.

In the course of the investigation, I retained the services of a professional electrical engineer and secured the use of a Westinghouse brush electrical meter where the individual electrical cycles could be analyzed. Working as a team, we isolated and analyzed the problem. An electrical current instability was measured from the compressor motor that affected that particular fluorescent lighting configuration and caused the flicker. The motor rotor mass and small diameter were such that the motor did not operate smoothly. A change to a compressor motor with a larger diameter and mass solved the problem. Neither the mechanical engineer nor the electrical engineer, operating individually, would likely have isolated the problem *and* found the solution.

The chemical engineer also plays a supporting role. Frequently, chemical reactions occur between materials or within fluids or lubricants that are beyond the expertise of the mechanical engineer and are implicated in a system failure. This is particularly true in failures involving laboratories or industrial ventilation systems. In these cases, the environmental system may be used to exhaust fumes or contaminants generated by chemical processes. The chemical engineer is then an equal partner in the investigation. The mechanical system may be the conduit but it is the chemists' materials that are being moved. The chemical engineer is the authority for the safe mixing, transportation, and disposal of these materials.

As the complexity of society increases, the industrial hygienist becomes an increasingly indispensable consultant because of two primary evolving phenomena:

1. As the use of chemicals in building construction and in daily life expands, the environment is being assaulted by ever-increasing concentrations of contaminants. The industrial hygienist identifies the problem. The environmental engineer solves the problem.

2. The energy standards mandated since the oil crisis of 1973 have resulted in tighter buildings and lower ventilation rates. Contaminants that used to be flushed from our spaces by excess ventilation are now concentrated indoors, causing a whole new set of problems. The "tight building syndrome" is the most well known. Working together, the industrial hygienist and the engineer are establishing new ventilation standards.

The mechanical engineer relies on the industrial hygienist to set the standards for acceptable exposure and to measure the existing concentrations in environmentally conditioned spaces.

Ambient noise and vibration are frequent participants in systems failures or complaints. In these cases, the acoustical consultant is the supporting authority. Although the mechanical engineer typically has a background in mechanical system noise and vibration control, which is supported by the *ASHRAE Handbooks*, expert analysis and testimony generally originate from the acoustical consultant.

Similar relationships exist between the mechanical engineer and other members of the engineering community, such as structural engineers, operating engineers, architects, and fire protection engineers.

A role not to be overlooked is that of the mechanical engineer as an expert consultant to each of the preceding disciplines, both in environmental systems failures and in failures in other fields. The mechanical engineer is only one member of a larger team of engineers and scientists constituting the field of forensic engineering.

9.6 CHANGE IN THE INDUSTRY

The rate at which products or design standards change in response to systems failures is directly proportional to the pressure placed on the producing parties. That pressure can be economic or legislative. The engineer sometimes despairs at the slow response to an obvious defect in a product; however, with the right pressure, mountains can be moved.

In the late 1950s, the State of California Office of Architecture and Construction refused to accept a class of two-position thermostats marketed by a major temperature control manufacturer. A potential safety hazard existed when a foreign metal object was inserted into the mechanism through the ventilation slots in the case. A California state agency recognized this potential liability and would not accept the product. Faced with the loss of a major market, the complete line of thermostats was immediately redesigned and replaced. Parts of that redesigned line are being sold and installed today. Without the pressure from a major user, that change would not have occurred for years.

The more frequent course is a slow and protracted evolution in improvement under pressure from the field engineers faced with systems failures. The more rapid and spectacular improvements occur under competitive and direct economic factors. Litigation and the threat of litigation can have a much more dramatic influence on change.

Prior to the energy crisis of the early 1970s, the ventilation standards for buildings were relatively loose. Many buildings were designed under the then-empirical 25% outside air rule. *ASHRAE Standard 62-73* recommended very conservative outside ventilation rates for various occupancies. For example, recommended office building ventilation rates were 15 to 25 CFM per person.

With the rapid rise in fuel prices and the related pressure for energy conservation in buildings, the ventilation rates were lowered to a minimum of 5 CFM per person. The result was the "tight building syndrome" and unacceptable indoor air pollution levels. Occupant complaints and litigation quickly followed.

Under this pressure, the air conditioning industry has responded. Minimum recommended ventilation rates have returned to much more realistic levels. The current ASHRAE ventilation standards recommend 20 CFM per person. It is now recognized that occupant health is more important than the cost of fuel.

An excellent example of how forensic activity has influenced environmental systems is the role that Legionnaires' disease has played in air conditioning system design. Prior to the initial well-publicized isolation of the *Legionella pneumophila* bacterium in a cooling tower, the design mechanical engineer had no concept of the implications of the cooling tower's location with respect to outside air intakes. Now, the engineer evaluates not only the tower location, but all other sources of potential contamination from the domestic and hydronic systems.

Back-flow double-check valves between the domestic water systems and the air conditioning hydronic systems have been recommended for years and are required by code. They usually were included in the system design and were generally, but not always, installed. Now they are *always* included in design and the engineer makes *sure* that they are installed and working properly.

9.7 CASE HISTORIES

9.7.1 44-Story Office Building, San Francisco

A 44-story high-rise office building in downtown San Francisco, California, was completed and occupied in 1973/1974 (Figure 9.1).

Figure 9.1 Forty-four-story office building with adjacent 22-story building, San Francisco.

9.7.1.a Complaint

The owner complained of excessive overheating, particularly in the upper floors. For example, with full cooling available and all heat shut off, the temperature on the 39th floor in early July 1976 was 90°F. The problem had to be resolved or litigation would result.

The design mechanical engineer, through periodic visits, final inspection, and investigation, had not isolated the problem. He suspected that the mechanical contractor had conducted an inadequate building startup, test, and balance, and that the conditioning zone terminal boxes did not conform to the specifications.

9.7.1.b System

The building is fully air conditioned. Cooling is provided by two 800-ton absorption chillers located in the lower basement that supply chilled water through vertical risers to fan rooms on the 2nd and 41st floors.

Heating is provided by building hot water generated from a steam heat exchanger in the lower basement machine room. Hot water rises vertically in parallel with the chilled water to the fan rooms. Additionally, each floor has a hot water loop distributing to a valveless heating coil at each air supply induction box on the exterior zones, one at each window, totaling over 2000 for the building.

The building heating hot water temperature is controlled by a three-way control valve reset by outside air temperature. When the outside temperature is above 70°F, this valve is in the full bypass position and supplies no additional heat to the system. Also, the circulating pump is turned off.

A parallel domestic hot water heating system shares the building steam heat exchanger with parallel risers to high-temperature hot water to domestic hot water heat exchangers on the 43rd floor. The temperature in this system is uncontrolled at 250°F with constant pump operation.

The system diagram (Figure 9.2) shows these related components and systems.

9.7.1.c Problem

The building heating system was consistently out of control, resulting in excessive building space heat. This overheating could not be corrected even by shutting down the building heating hot water pumps. Nothing seemed to be able to eliminate the excessive overheating. The overheating problem tended to be worst in the higher floors. The building heating temperature control valve at the basement heat exchanger was controlling correctly.

With the building heating pump shut off and the check/shutoff valve closed, there was still hot water in the individual floor hot water distribution loop. Heating continued to occur although all the space thermostats were calling for cooling.

Attention had been focused on the individual temperature control induction terminal boxes and their operation and compliance with the specifications. The function of these boxes is to take primary cooling air from the air distribution system through a variable air volume cycle to satisfy space cooling. To satisfy heating needs, an induction nozzle in the primary airstream induces warm return air from the attic through heating control dampers and into the space conditioning airstream, as primary cooling air is throttled to a minimum of 50%. In the perimeter induction boxes, an additional heating coil in the induced airstream supplements the attic air for heating. This coil gets its hot water from a building heating hot water loop at each floor served from the heat exchangers in the basement. The coil did not have a temperature control valve. To limit heating, the space thermostat controlled a set of dampers in the induced airstream

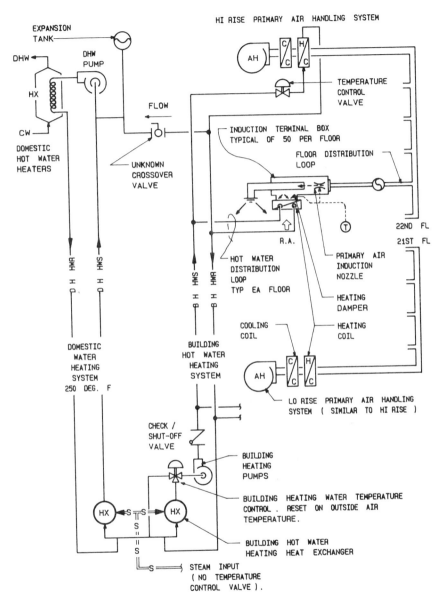

Figure 9.2 Heating system air distribution system schematic for 44-story office building in San Francisco. The unknown crossover pipe and valve that caused the primary environmental system failure are shown. This pipe allowed high-temperature water to flow from the building heat exchangers through the building heating system, returning through the domestic hot water system.

between the heating coil and the induction nozzle which modulated closed when the thermostat was satisfied on heat. These dampers were specified to be "tight closing" (Figure 9.3).

The prevailing question was: If the building heating hot water system was shut down and the induction dampers were closed, how could the building still be heating?

Concentrating on the symptoms, the owner was prepared to issue a $250,000 (1976 dollars) contract to install a temperature control valve on each of the over 2000 induction box heating coils to stop positively the flow of heating water. This solution would have also placed a dollar value on a potential claim against the architect. Added to this cost was the value of the unoccupiable space while the problem existed. The potential for a large liability claim existed.

9.7.1.d *Investigation*

The author was retained by the architect to find the cause of the problem and to make recommendations for its correction. This commission was held to the highest level of professionalism in that the interests of the owner and truth were put ahead of possible damages to the architect.

The detailed survey and analysis of the HVAC system looked into all facets of the heating systems and related cooling and domestic hot water systems. The focus of the investigation was determination of the root cause or causes of the problem. As it turned out, there existed several contributing factors in addition to the one very simple, basic cause.

Numerous systems and subsystems could have had an impact on the problem: building heating water flow, induction box performance, equipment and temperature control, solar loading on the structure, building stack effect and leakage, individual component operation and failures, domestic hot water heating system, and so on. All of these interplay and impact on the total building system.

To establish order, each system was compartmentalized and then subcompartmentalized, with complete documentation of each maintained. Essentially, the building was a laboratory and the individual experiments were run as if they were academic laboratory projects. Throughout the compartmentalization, the interrelationship of the various systems had to be considered and analyzed.

A series of experiments was set up to answer the following questions:

1. *What was the flow in the various hot water risers under supposedly static, not pumping, conditions?* Because of the stack effect of the building with a prime source of heat in the basement and without positive control of the steam supply, a tendency for gravity flow in the return piping was discovered even when the supply piping was shut off.

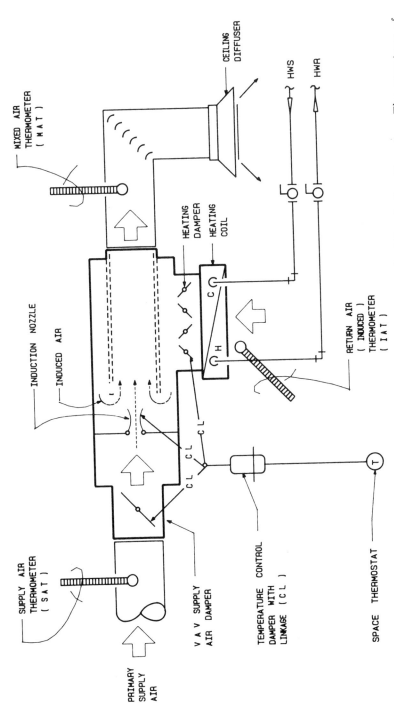

Figure 9.3 Induction box schematic with measuring points for 44-story office building in San Francisco. The symptoms of the failure were experienced here. To quantify the effect of return air leaking through the heating damper at very low air flow rates, thermometers were used to measure the return air, supply air, and mixed air temperatures. These measurements were used to calculate the percentage leakage through the heating damper.

2. *How could heat enter the space through the induction boxes on a call for cooling and with the induction heating damper closed?* It was found that the induction dampers leaked as much as 30% (averaging 15%) of the total supply air past the blades (when closed) and the induction nozzle at full cooling. The specifications required that these dampers be "tight closing," but there were no quantifying values to identify what "tight closing" meant. Consequently, on an industry-wide standard at that time, this leakage was not considered significant. The linkage did close "tight."

A laboratory environment was set up on a statistical sampling of a group of induction boxes. The air leaking through the dampers and thus through the heating coil moved at a very low velocity—much too low for field air flow measurement. Therefore, the leakage was measured using accurate air temperature instruments. Sixteen-inch-long glass thermometers accurate to 0.1°F were used to measure the primary supply air temperature, the attic air just before it entered the heating coil, and the mixed air temperatures. Each set of measurements was repeated several times to prove repeatability and to "close" and make sure that conditions had not changed during the set of readings.

On the basis of these measurements, it was concluded that, with the dampers in the full closed position, warm attic air was leaking past the damper blades, was heated significantly as it passed over the hot heating coil, and raised the supply air temperature to the space sufficiently so that space overheating occurred.

If the heating coil had not been "hot" during a normal cooling cycle, this overheating would not have occurred because the boxes would have been operating within their safety factor. The root cause of the problem was the hot water in the building hot water system and heating coil that should have been shut off during a normal cooling cycle.

3. *Where did the hot water used by the building heating system when that system's pumps were shut off and the check/shutoff valve was closed originate?* The only likely source of uncontrolled hot water was the adjacent domestic hot water heating system. The two systems did share a common return water loop, but no other crossovers were shown on the plans or were obvious from inspection.

These are times when an ESFE must be prepared to get dirty and to crawl through the building. The ESFE must throw the suit and tie away, put on the coveralls, and go looking.

A small 1-inch-diameter crossover pipe was found

near the ceiling of the penthouse equipment room connecting the domestic hot water heating water system to the building heating water system through the expansion tank piping. Installed in this crossover pipe was a 1-inch-diameter ball valve left in the "open" position. The crossover pipe and valve were not shown on the plans. Very likely, the pipe was installed by the contractor to facilitate the air purge when filling the system, and then it was left open and forgotten. Of course, that is mere speculation. How it got there was irrelevant. The point was that it was there.

The building problem was immediately solved by closing the ball valve.

Other problems were discovered during the investigation, some of which contributed to or exacerbated the foregoing problem:

1. The building temperature control indicating system provided readings that were inaccurate. These readings made it very difficult to isolate a problem or believe what was happening. This item was corrected by the contractor under warranty provisions.
2. A series of equipment failures, correctable under warranty, were discovered during the investigation. These were normal and of little final consequence. During the investigation, however, they tended to lead into blind alleys and to give false information. If these had not been completely traced, entirely different and expedient conclusions may have been made that would have masked the true root cause.
3. The building is subject to a significant "stack" effect. All high-rise buildings are subject to this effect where cool air tends to enter through the lower sections and warm air exhausts at the top. An exterior panel sealant failure exaggerated the stack effect. This had a tendency to chase the symptoms of the overheating problem up and down the building in accordance with the outside winds and temperature. This, in conjunction with inadequate building static pressure controllers, wasted numerous investigation hours. The false indications caused by these effects led the author on many a merry (but false) chase.
4. During prior attempts to correct the problem, the heating system hydronic balance had been severely altered, leading to false clues.

9.7.1.e Conclusions

The root problem was caused by an obscure crossover pipe between the building heating system and the domestic hot water heating system that induced 200°-plus building heating system return water up the return riser, across the bypass at the expansion valve intertie, into the domestic hot water heating system at the domestic hot water pump suction. The induced reverse flow was intensified by the heating system's tendency to sustain gravity flow throughout the high-rise system. The problem was solved by closing the 1-inch-diameter ball valve in this crossover.

The closing of the ball valve (a 10-second task once the valve was found) allowed the building to work as designed, alleviated the need for a $250,000 expense to install control valves at each induction box, and relieved the architect and design mechanical engineer of a potentially expensive liability.

When searching for the root cause and thus the truth in an environmental systems failure, the manifest symptoms may only be the result of the failure, not the cause. Neglecting to pursue to a final conclusion can lead to a false conclusion and an injustice.

The ESFE should expect false indications and misleading information and should not be deterred by other factors. There seems to be a correlation between the effort expended and the depth of truth finally discovered.

9.7.2 Four-Story Motel Building, Lake Tahoe, California

This study demonstrates how the ESFE may protect the client from making a serious or costly mistake through unsubstantiated litigation. One of the services the ESFE must provide is to evaluate and challenge a situation as early as possible so the client knows how valid his or her position is before pursuing a claim or litigation.

9.7.2.a *Complaint*

A contractor who had installed a 50-ton package chiller on the roof of a motel was sure that he had been sold a "used" chiller instead of a new unit. His suspicions were based on the condition of the chiller when it arrived at the job site and on a series of equipment failures during the first year after startup.

He was in the process of preparing to file a suit against the chiller manufacturer for selling a used unit in place of the new unit he had ordered. I was retained by the contractor to investigate the unit. His intent was to prove that the chiller was a used piece of equipment. His claim was supported by his attorney who had written a letter to the manufacturer, complaining of problems with the unit.

9.7.2.b *System*

The contractor had multiple licenses—general contractor, mechanical contractor, and plumbing contractor. Using these licenses, he had designed the HVAC system and installed both the plumbing and the HVAC systems while doing the general construction work.

To design the HVAC, he had relied heavily on equipment vendors to calculate the building loads and to size the equipment. Neither the contractor nor the equipment vendors were professional engineers.

The HVAC system was a four-pipe hot water heating, chilled water cooling system with two boilers in the main-floor boiler room, the 50-ton chiller mounted on the roof, and fan coil units installed in each motel room.

9.7.2.c Problem

The package air-cooled chiller had arrived by truck at the job site with physical damage to the frame (see Figure 9.4). The contractor had formed his suspicions about the unit being used on the basis of this damage and a series of observations regarding features of the delivered unit:

1. Items ordered with the unit were not installed and had to be field installed by the manufacturer's representative. The contractor felt that these items should have been factory-installed on a new unit.
2. The starting contacts on the compressor starters showed evidence of arcing, as if the unit had been run before.
3. The compressors were produced by a different manufacturer from the basic chiller unit. As the chiller manufacturer also produced compressors, the contractor suspected that the compressors had been replaced in an old chiller assembly.
4. During field startup, refrigerant oil was found in the evaporator chiller barrel. The oil could have been there only if the unit had been previously run. Also, this oil was black and smelled burnt.
5. On startup, errors were found in the starting controls that required correction before the unit could be run.
6. The unit experienced an excessive series of condenser motor failures during the first year of operation. The contractor felt that these failures would not have occurred in a new unit.
7. The shipper's bill of lading showed that the unit came through Florida on its way to Lake Tahoe. The unit was manufactured in Pennsylvania.

9.7.2.d Investigation

My field investigation confirmed the contractor's observations. The unit was in very poor condition and should not have suffered the damage and failures that had occurred. To validate the observations and suspicions, I investigated each item, attempting to verify all conclusions through a second source. More than anything, this was a paper trail.

Figure 9.4 Four-story motel building chiller, Lake Tahoe, California. Examples of physical damage to the unit as delivered to the job site.

The following information was collected:

1. It is common industry practice to field-install items that have been special-ordered to replace stock items. In this case, the field installation was poorly done. Although this did not validate the opinion that it was a used unit, it did warrant

Figure 9.5 Four-story motel building chiller, Lake Tahoe, California. Examples of poor installation work performed by the manufacturer's field mechanics to fulfill unit purchase requirements.

a claim to correct the unacceptable workmanship (see Figure 9.5).

2. The history of manufacture was obtained from a source other than the original seller. These dates presented a pattern of manufacture, transit, and delivery that did not include a significant time period that would have allowed the unit to have been used elsewhere. To support this history, the dates of manufacture of the two 25-ton compressors coincided with the history of the chiller unit.

3. The unit had been test-run at the factory. To do this without a load, some of the operating controls had to be bypassed. This test run could account for the arced electrical starter contacts and oil in the evaporator barrel. The test run could also account for the evidence that the unit had been previously run. However, it would explain the black and burnt oil only if there had been a problem on the test run. This was a possibility.

4. The manufacturer's catalogs suggested that compressors from a different manufacturer were normally supplied with this size unit.

5. The condenser fan failures were excessive and inexcusable; however, these problems were not necessarily the result of a used piece of equipment. It was much more likely that these failures occurred because the fan motors were started while they were free-wheeling backward under the reverse air flow of adjacent fans. Starting under these conditions can cause a severe overcurrent condition, which would be exacerbated by the poor motor cooling capabilities of air at 6200 feet elevation.

9.7.2.e Conclusions

No irrefutable evidence was found to prove that this unit was, in fact, "used." All the circumstantial evidence could be explained by other conditions. The manufacturer could present evidence substantiating the date of manufacture of the unit, as well as the date of manufacture or delivery of component parts. It could not be proved, or for that matter conclusively disproved, that the unit was new, as represented by the manufacturer and supported by the preponderance of the evidence. A claim for a defective unit could have been substantiated, but a claim for a used unit could not.

My report to the client advised him of the reality of the situation as I had evaluated it, with a recommendation that he not pursue a legal action based primarily on emotion. He had been harmed by the condition of the chiller unit and by the lack of support provided by the manufacturer. To seek recourse, however, he needed to have a particularly strong case. A weak case that would likely end in failure would only cost him more money and compound his damages.

After my investigation and report, the contractor's legal actions against the manufacturer were terminated.

9.7.3 Three-Story County Administration Building, Central California

This study demonstrates that numerous minor failures, each individually insignificant, can combine to cause a major environmental systems failure and an unsatisfactory building.

9.7.3.a Complaint

The owner of a recently completed office building complained of unacceptable environmental conditions:

Unsatisfactory space temperature

Poor outside air economizer operation

Inadequate air supply to areas located at the extremities of the ductwork.

Poor return air from conference rooms and from the outlying portions of the building.

Poor and inconsistent temperature control operation.

Higher-than-projected energy consumption

He retained my firm to investigate the problem, to find the cause and to identify the responsible party so that a claim could be made and the problem corrected.

9.7.3.b System

The building's HVAC system primary equipment is located in a penthouse fan room. There is a primary air conditioning air handler system that serves most of the building, as well as three additional small air handling systems for special areas. Most of the complaints seemed to relate to the primary, built-up air handler system.

Heating and cooling are provided from a conventional hot water boiler and a set of two roof-mounted air-cooled chillers. Hot water and chilled water are delivered to the heating and cooling coils through a four-pipe distribution system with a pump set in each loop.

The air distribution system from the primary air handler system runs from the supply blower, across the fan room, down to each of the three floors, and is distributed to the various areas on each floor. Individual space control is by variable air volume (VAV) boxes controlled by a space thermostat for each box. Return air is ducted from each area directly back to the fan room to a return air fan and to a conventional outside air–exhaust air mixed air economizer system. Sound traps in the duct are located in the supply air from the blower to the building and in the return air before the return air blower.

The temperature control system receives input from the various space thermostats controlling the VAV boxes and operates the heating and cooling valves as well as the outside air economizer through a discriminator circuit. This system is very complex.

9.7.3.c Investigation

Construction plans, specifications, and other contract documents were studied to identify what the system was designed to do. A thorough site investigation was conducted (1) to quantify actual operating conditions for comparison with contract requirements and (2) to identify the detailed operating characteristics of the system. As poor space air quality was the focus of many of the complaints, this subsystem was investigated first.

A set of on-site air flow measurements was made based on a gross measurement at the filter bank where the total air quantity was stable. This measurement was made with the air handling system operating at full available volume. A statistical set of space supply air measurements was obtained to identify those areas that had the lowest flow rates. It was found that the poorest air distribution was on the second floor at the furthest duct run.

While operating at the measured air quantity, duct static pressure readings were taken throughout the duct length. These readings very quickly identified the trouble spots where excessive pressure losses occurred. It was concluded that the cause of the poor air quality was low air quantities caused by excessive pressure losses in the duct system.

Air flow calculations were made to correlate the measured values with theoretical values for evaluation of system performance. This led to detailed calculations of some of the individual elements within the system.

The temperature control system was checked for operation. This system was substituted by the mechanical contractor from the specified system under the approved terms of the "or equal" clause in the specifications. As mentioned previously, the system was very complex. Although it could perform the task by design, in actual practice it was not operating satisfactorily. This was compounded by the building operating engineers' lack of understanding of and confidence in the equipment. The resulting poor temperature control operation caused the excessive building energy consumption.

Although the poor air quality in the building was traced to the air distribution system, the problem was the sum of a series of individual subsystem failures:

1. The mechanical engineer had designed the system under the then-current energy conservation standards which allowed little room for error. His system was also constrained by a tight construction budget and duct location limitations in the architectural design. Too small a safety factor was built in to account for changing installation conditions.

2. During construction a conflict was discovered between the stairs leading into the fan room and the well-designed sweep elbow in the ductwork at the first turn from the blower discharge. To resolve this, the mechanical contractor elected to install a square elbow directly at the discharge from the blower. This configuration can easily result in a 30% system loss factor over the designed configuration. To compound the problem, the contractor installed square-corner turning vanes instead of the specified double-width curved vanes.

3. An excessive pressure drop was discovered in a series of square duct elbows where the engineer had to route the ductwork around an elevator shaft. This was the result of an architectural constraint. Further investigation of this problem (accomplished by crawling inside the duct) disclosed that although the specified type of duct turning vanes had been

installed, they had been installed in such a way that more pressure was lost than had been calculated.

4. Excessive pressure loss was also discovered in a bull-head duct tee where the supply duct dropped from the fan room into the second floor ceiling. Here, the duct came straight down to a capped end. Two horizontal side ducts took off in opposite directions to different areas of that floor. These takeoffs had neither turning vanes nor splitters (see Figure 9.6).

5. The VAV control inlet vanes to the primary supply blower were operating at such a point on the blower fan curves that

Figure 9.6 Three-story county administration building, central California. The bull-head tee without turning vanes and poor turning vanes in the associated ductwork caused an excessive static pressure loss. The detail shows the recommended turning vane installation.

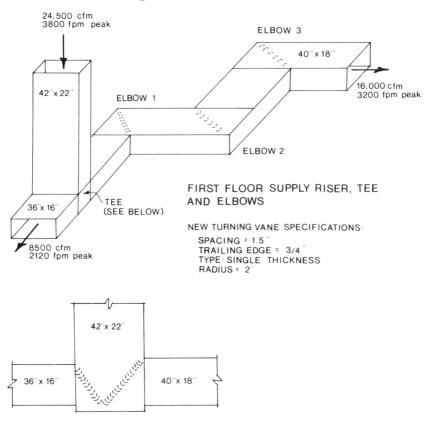

they had little or no effect. Consequently, the blower was operating essentially as a constant-volume device. This was a result of the various system effects that displaced the air distribution system from its calculated rating.

6. The same types of system and installation problems existing in the supply air system were found in the return air system.

9.7.3.d *Conclusions*

There was an extensive list of problems, each one individually not serious enough to cause a significant impact on system performance; however, when taken as a whole, these deficiencies caused a major building environmental systems failure. And everyone participating in the project participated in creating the problems:

Owner. Severe economic constraints influenced the type of system designed. The owner's operating personnel were not sufficiently experienced or trained to maintain a building of this complexity. This was particularly evident with the controls.

Architect. Because of the positioning of some building elements, the mechanical engineer had to route the air distribution system in such a manner that excessive duct static pressure losses were inevitable. A conflict in the plans caused the installing mechanical contractor to reroute the duct just downstream from the supply blower discharge.

Mechanical Engineer. Not a large enough safety factor was allowed to take care of the several job conditions that inevitably develop during construction.

Installing Mechanical Contractor. An inappropriate solution was applied to the supply blower discharge when the conflict with the stairs was discovered. This resulted in a very large system loss which was compounded by installation of the wrong type of turning vane. The contractor's mechanics did a less-than-satisfactory job of installing the turning vanes in various duct elbows.

Because of a lack of detail at the duct down-comer to the second floor ceiling, the mechanical contractor installed an inappropriate bull-head tee. If he had consulted with the engineer at that time, this problem could have been avoided.

Equipment Suppliers. The temperature control contractor designed and installed a control system so complex that it could not be accurately calibrated. It was not set up to function satisfactorily. Granted, the system could theoretically do anything. From a practical standpoint, however, it was not an appropriate system. The system was simply too complex for the owner's operating engineers to comprehend.

As part of our report, recommendations were made for corrective action. These took several paths:

1. The mechanical contractor was to correct the blower discharge elbow, bull-head tee, duct elbow turning vanes, and several other lesser problems that were discovered.
2. The basic inadequate system static pressure was relieved by removing installed sound traps and the inherent pressure loss the traps cause. Fortunately, the system was quieter than anticipated and these sound traps were not as important as originally expected (see Figure 9.7).
3. The temperature controls were to be simplified so they could be expected to operate satisfactorily.

Figure 9.7 Three-story county administration building, central California. The fan room plan shows the configuration of the severe elbow directly downstream from the supply air fan primary blower. Also shown is the location of the sound traps which were removed to reduce the system static pressure loss.

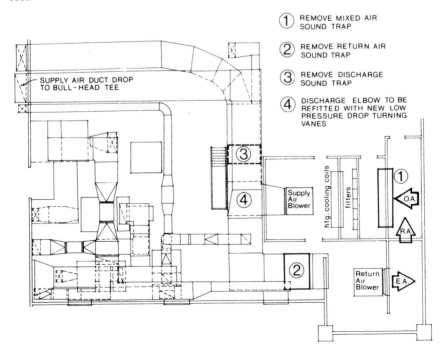

4. The building owner was to establish an extensive training program for the operating engineers and mechanics.

And finally, who was at fault, contributing to the resultant environmental system failure? Everybody.

9.8 CONCLUSIONS

Environmental systems, in building, industrial, process, or microclimate applications, are subject to failure. Each environmental system comprises a myriad of subsystems, any one of which can fail and bring down the total system. As long as we place so much pressure on our environmental systems, failures will continue to occur, and the services of the environmental systems forensic engineer will continue to be needed. Fortunately, these failures are seldom catastrophic, where life and limbs are at risk. They can, however, make an environment uninhabitable or unbearable, resulting in great discomfort and monetary loss. Whether a failure is catastrophic or minor, the ESFE's job is the same—to accomplish the work with accuracy and thoroughness that can stand the test of scrutiny and challenge.

The ESFE is a diagnostician, detective, reporter, and "expert." He or she may also be a corrective engineer. All of these positions require, as a foundation, an extensive command of physics, processes, systems, equipment, details, requirements, and related disciplines for each system under study. Also required are analytical skills: the ability to observe a large mass of impressions, realize their individual and collective significance, evaluate them in the light of physics and psychology, and arrive at a rational conclusion. As no single engineer can be expected to have this extensive a command, the ESFE must have the support of experts in other engineering disciplines. The range of expertise needed in evaluating environmental systems failures is immense.

Irrespective of who the client is, all work must be conducted with integrity. The opportunities for conflict of interest are many and only the highest standards of personal and corporate integrity will suffice.

At the outset, the ESFE is a detective. He or she must have good investigative techniques. The investigator usually starts by interrogating the client and other affected parties to identify the true problem. This is normally followed by an extensive analysis of the failure, consultation with supporting disciplines, development of conclusions, and preparation of the report. Court proceedings or, if no litigation is required, corrective engineering may follow.

At this juncture, the ESFE is faced with an ethical dilemma: The ESFE is retained to seek the truth in a situation and report it to the client. The ESFE cannot honestly perform that service if he or she is simultaneously soliciting the design contract for corrective engineering. Yet the ESFE should know the corrective solutions if the report is to be considered valid. After resolution of the initial commission and the ESFE's report, if independently approached and with full agreement of all affected parties, the ESFE may undertake additional corrective engineering services which he or she may be uniquely qualified to perform. To maintain integrity, however, the official report must be prepared so that any qualified engineer could implement the correction.

The ESFE's tools establish the accuracy of the findings. The field instruments should be as basic as possible and inherently accurate with a traceable calibration history. The analytical tools of engineering must be founded in basic physics principles and recognized standards, such as the ASHRAE handbooks.

The practice of environmental systems forensic engineering can be stimulating from an intellectual standpoint. It requires extensive background and experience. At the same time, forensic practice is sobering with the realization of the gravity of the situation. The work and conclusions of the ESFE can have consequential impact on people and economics. Therefore, this work must be thorough, accurate, and just and must be prepared with faultless integrity.

References

American Conference of Governmental Industrial Hygienists (ACGIH). 1986. *Industrial Ventilation 19th Edition. A Manual of Recommended Practices.* Cincinnati, OH: ACGIH.

American Society of Heating, Refrigerating and Air Conditioning Engineers Inc. (ASHRAE). *Equipment Volume, Fundamentals Volume, Refrigeration Volume, HVAC Systems and Applications,* current volumes (one volume published annually). Atlanta, GA: ASHRAE.

———— 1981. *ASHRAE Standard 62-1981: Ventilation for Acceptable Indoor Air Quality.* Atlanta, GA: ASHRAE.

———— *ASHRAE Transactions* (technical and symposium papers, published annually). Atlanta, GA: ASHRAE.

Associated Air Balance Council (AABC). 1982. *National Standards for Total System Balance* 4th ed. Washington, DC: AABC.

System Design Manual, all eds. Syracuse, NY: Carrier Air Conditioning Company.

Trane Air Conditioning Manual, all eds. La Crosse, WI: The Trane Company.

10. Forensic Photogrammetry

WILLIAM G. HYZER, P.E.

10.1 INTRODUCTION

It is difficult to imagine any civil or criminal case requiring the services of an engineering expert that does not include the use of photographs in development of the case and in support of testimony. Pictures are far more effective than words in describing almost any object or scene. Take a common object, a bolt for example, and try to describe it to someone who has never seen one. Few words are required to describe the detrimental effect of a particle caught between the contacts of an electrical relay, when a single photograph illustrates it so well (Figure 10.1). Graphic representations, of which this photograph is only one example, are very important tools of communication for the expert in dealing with lawyers, jurors, and other experts.

The value of ordinary photography to forensic science and criminalistics has been well documented (Eastman Kodak 7-71,10-72). *Forensic science* is simply defined here as the application of those portions of all sciences as they relate to the law. *Criminalistics* is defined as that profession and scientific discipline directed to the recognition, identification, individualization, and interpretation of physical evidence through application of the natural sciences and science matters (Dragel 1965). Most attorneys recognize the importance of photography in these matters, but few utilize it to full advantage, particularly in those cases that involve technical judgments and decisions.

My purpose in writing this chapter is to point out some of the specialized photographic techniques that could be more effectively

Figure 10.1 Photomacrograph of a particle caught between electrical contacts.

applied in forensic and criminalistic matters. The ideas presented here are based for the most part on my own personal experience as a scientific consultant in the fields to be mentioned.

The term *photography* is interpreted in this chapter as any direct imaging process using a camera, including the video camera. Photography, therefore, is inclusive of videography, instant photography, and conventional photography using black-and-white and color films. *Photogrammetry* is the art or process of surveying or mapping with the help of photographs.

The primary objective in most forensic applications of photography is to produce images that accurately represent the object, scene, or event depicted. Images that are subject to more than one interpretation or that misrepresent reality create contradictions that can do more damage than good. This chapter deals primarily with techniques for obtaining accurate and informative photographic records for legal applications.

Forensic photographs can be divided into two broad categories: *documentary* and *metric*. A documentary photograph is analogous to a rough sketch, drawn to no particular proportion or scale. A metric photograph, on the other hand, is comparable to an engineering drawing or map because it contains the information necessary for deriving quantitative spatial data.

Product liability cases often hinge on technical matters that can easily pass "over the heads" of many jurors. Well-planned photographic evidence effectively narrows the technological gap between the expert testimony of a knowledgeable witness and the understanding of the ordinary juror. It should be noted that oral testimony may be illustrated with photographs of actual objects or scenes similar in classification and appearance to the original, but photographs of objects or scenes artificially prepared are usually not admissible. Also, ordinary photographs of an object are usually not admissible if the original object is before or accessible to the jury.

10.2 PHOTOGRAPHIC TESTIMONY

The Defense Research Institute, a nonprofit legal organization that maintains a file of expert witnesses, defines an *expert* as "a person with knowledge in a given field gained through education and/or experience, with sufficient technical ability, and with the ability to articulate on his/her knowledge so as to place him/her in a position as somewhat of an authority on the subject, beyond that of the layperson" (Jacobson 1973). The test of admissibility of expert testimony in court is "whether the witness' particular skill and knowledge, not common to persons in general, enables the witness to draw an inference, where persons of common experience with all the facts proved would be left in doubt" (*New England Glass v. Lovell, 7 Cush. Mass. 319*)

Although the practicing forensic engineer may be an expert in structural engineering, fracture mechanics, safety engineering, or any other field of specialty, care should be exercised to avoid being qualified as an expert in photography. Such qualification is an open invitation to the opposing attorney to inquire into the witness' knowledge of photographic chemistry, lens design, and other highly technical areas of photographic science. Instead, the engineering expert should state that he or she is a serious practitioner of photography within the field of expertise, but not a photographic expert. Photography is simply another tool to aid the witness in investigative work. If the witness has specialized training or other qualifications as a photographic practitioner, these are worth mentioning, but a witness should be careful not to stray outside his or her limits of competence.

10.3 PHOTOGRAPHIC TECHNIQUES

The many techniques of photography that deserve consideration in technologically oriented cases are summarized in Table 10.1. The table includes a brief description of each photographic method and representative applications (Hyzer 1975).

10.3.1 Cameras

A wide variety of cameras have application in forensic photography, but the 35-mm single-lens reflex (SLR) camera is the most versatile

Table 10.1 Specialized Imaging Methods and Related Techniques Having Forensic Application*

Technique	Brief Description	Typical Forensic Application
High-speed cinematography	Motion pictures recorded at supernormal rates	To reveal detailed actions in a high-speed process or event that the eye cannot ordinarily follow, such as gear chatter in a high-speed machine, splashing of liquid, and deflection of a reciprocating saw blade
High-speed photography	The use of short-duration light sources or high-speed shutters to record rapidly moving objects with minimal image blur	To reveal detailed pictorial information at discrete stages in the action of rapidly moving phenomena, exemplified by a bullet in flight, fracture of a glass plate, implosion of a vacuum tube, etc.
Holography	A three-dimensional imaging process utilizing two beams of coherent light to produce an interference pattern on a light-sensitive emulsion	To provide a three-dimensional optical model of an object
Infrared photography	Imaging by radiation in the near-infrared region of the spectrum	Recording scenes illuminated by invisible infrared radiation as in surveillance, or utilizing infrared radiation to reveal information that is invisible in ordinary light, such as printing on a charred document
Instant replay	Electronic imaging by video or television processes on magnetic tape or disk for immediate playback on a TV monitor	To record an action for immediate analysis, usually by slow-motion or stop-action techniques
In-water photography	Photography performed beneath the surface of water with cameras in special waterproof housings	Recording evidence at the bottom of lakes, rivers, harbors, etc.

(continued)

Table 10.1 *(continued)*

Technique	Brief Description	Typical Forensic Application
Laser micrography	In situ photographic or video recording of fluid-borne, micrometer-size particles	To reveal the presence and size of harmful particles, such as asbestos fibers, toxic dusts, and poisonous aerosols, in the air
Low-Light-level photography	Recording scenes or objects with existing light at very low ambient levels, by means of conventional photography using highly sensitive films or by electronic image intensification	To permit photography under nighttime conditions without supplementary illumination as in surveillance applications, recording of glowing objects, etc.
Optical photomicrography	Photography performed through an optical compound microscope at optical magnifications ordinarily ranging from 10:1 to 1000:1	To provide enlarged views of extremely small specimen areas such as corroded surfaces, wear marks, and dust particles
Photoelastic recording	Photography of transparent modes in polarized light to reveal areas of stress	To determine the location and magnitude of stress in glass or plastic parts, such as bottles, or models of mechanical parts, such as gears, screws, brackets, and shafts
Photogrammetry	The science of making physical measurements by photographic methods	To determine the distance between objects or the size of objects recorded in a photograph, such as the distance between two automobiles in an accident scene and the depth of a crack recorded in a photomicrograph
Photomacrography	Recording of small specimens at optical magnifications ordinarily ranging from 1:1 to about 30:1	To provide enlarged views of small specimens such as hairs, insects, electrical contacts, and fractured parts

Table 10.1 (continued)

Technique	Brief Description	Typical Forensic Application
Photometry and radiometry	The science of measuring radiant energy within the optical spectrum	Specifying the brightness or color of an object or the level of illumination at a point in space, such as the brightness of a spark or the illumination produced by a flame
Scanning electron microscopy (SEM)	Electronic imaging performed by scanning the specimen with an extremely narrow beam of electrons to produce magnifications ranging upward to 100,000:1	To provide highly magnified views of the surface of a specimen to reveal such morphological details as height of surface features, chemical composition and phase determination
Schlieren photography	An optical method of recording gradients in the refractive index of transparent solids, liquids, and gases	To reveal dynamic flow lines and thermal gradients in fluids and optical inhomogeneities in transparent solids resulting from pressure, stress, localized heating, etc.
Stereophotography	A three-dimensional imaging process using a pair of spatially separated photographs to simulate binocular vision when viewed through a stereoscope or by stereo projection	To provide a three-dimensional representation of an object or scene
Telescopic photography	Photography through a telescope	To record objects at a great distance that cannot be conveniently approached as in surveillance photography and recording of dangerous objects or events
Thermography	Imaging of warm to hot objects, usually by electro-optical methods	To reveal variations in surface temperatures of objects such as furnaces, transmission lines, transformers, electronic devices, motors, and bearings

(continued)

Table 10.1 (*continued*)

Technique	Brief Description	Typical Forensic Application
Time-lapse cinematography	Motion pictures recorded at subnormal rates	To reveal detailed actions in a slowly changing process or event that cannot ordinarily be seen by direct observation, such as the rusting of iron, soil erosion, crystal growth, and aging of rubber
Ultraviolet photography	Imaging by radiation in the ultraviolet region of the spectrum	Utilizing ultraviolet light to reveal information that is invisible in ordinary light, such as pigmentation in certain paints and the presence of fluorescent dyes
X-ray photography	Imaging by means of penetrating X-rays or gamma radiation	To reveal conditions inside of an otherwise opaque object such as internal cracks, defective welds, and missing or improperly installed internal parts

ˑFrom Hyzer, W. G. 1975. Your day in court could be close at hand. *Photomethods* 18, No. 8.

and widely used for this purpose; 35-mm SLR cameras and techniques are emphasized in this chapter. Instant cameras of the type manufactured by Polaroid Corporation also have applications and are in common use today. Video cameras utilizing both $\frac{1}{2}$-inch and $\frac{3}{4}$-inch videotape are also finding applications in this field. There are advantages and disadvantages to each of these systems. A universal camera does not exist.

The 35-mm SLR camera offers accurate framing, easy focusing, interchangeable lenses, flexibility in choice of films, portability, and versatility in handling a broad variety of subjects under a wide range of lighting conditions. This camera system is capable of producing transparencies for projection or prints for direct viewing, of a quality that is unsurpassed by any other imaging system. The major disadvantage of wet-processing photography is the inability to view the image within a reasonable time after it is exposed.

The problem of delayed viewing of an image is solved with Polaroid cameras, but at the sacrifice of versatility and image quality. The same applies to video cameras. There are many instances, however, when it becomes extremely important to preview an image at the scene where it is recorded. There have been many cases where valuable evidence has been lost because of operator error or camera malfunction at the scene that was not discovered until after the film was processed. Instant photography and videography eliminate this possibility as long as the recordings themselves are not lost or mutilated after previewing. In those cases where a photographic record is vital or costly to repeat, either instant photography or videography might be considered as a backup for the conventional SLR camera. A logical approach in a typical case might be to first record the subject matter using instant photography. Polaroid print films and 35-mm transparency films in both black and white and color are available for this purpose. Careful examination of the prints or transparencies on site will then be helpful in selecting the proper vantage point, lighting, and exposure for subsequent photography with conventional film in an SLR camera. In this way, instant photography serves two useful purposes: (1) a means of previewing the image and (2) a backup to the SLR camera.

There is a tendency, especially among nonprofessionals, to place more emphasis on conserving film than on adequately recording a scene with a sufficient variety of angles and exposures to ensure proper coverage. The 35-mm film is relatively inexpensive compared with other factors in an expert's investigation of a legal case. Film should be exposed profusely.

10.3.2 SLR Camera Lenses

In addition to a 35-mm SLR camera body, an assortment of lenses and other photographic accessories are usually required to meet the investigator's specific needs. A 28- to 85-mm or 35- to 105-mm zoom lens fulfills most focal length requirements. The short-focal-length lens is useful for recording scenes in which the working distance is limited. The long-focal-length lens permits distant objects to be recorded that cannot otherwise be easily approached for close examination. Representative applications for these two focal-length settings with a zoom lens might be found in a typical fire-damaged building. The wide-angle setting might be used to record the overall damage in a room. The telephoto setting, on the other hand, might be useful in photographing the condition of an electrical junction box visible through a hole in the ceiling.

Although a zoom lens incorporating either of these ranges of focal length fulfills many, if not most, lens requirements, at least two other lenses are usually needed in the forensic expert's photographic kit. The maximum aperture or "speed" of a zoom lens is usually limited to f/3.5. A "faster" lens is often required under low-light-level conditions or when high shutter speeds are required to arrest subject motion.

Image distortion is also greater in most zoom lenses compared with a good quality fixed-focal-length lens. Distortion becomes important when a photograph is used as a basis for making spatial measurements of a recorded object or scene. A high-quality 50-mm f/1.2 or f/1.4 lens fulfills most "fast" lens requirements. There is a widespread misconception among attorneys and other laypersons that a 50-mm lens is a *normal* lens and is absolutely required to record a scene in proper perspective. The terms *perspective* and *distortion* are also erroneously interchanged by the misinformed. It is not uncommon practice for an attorney to attempt to disqualify a photograph as evidence because it was not recorded with a *normal* 50-mm lens and therefore the image is *distorted*. The ability to record a scene in true and accurate perspective is not limited to 50-mm lenses alone. Lenses of short and long focal lengths are also capable of recording a scene in true perspective and are usually free of distortion for all practical purposes when they are applied properly. Geometrically correct perspective is maintained in viewing the image of a scene when the viewing distance is $E \times F$, where E is the enlargement factor in producing the print or projected image and F is the lens focal distance. For example, an $8\times$ enlargement from a negative exposed through an 85-mm lens should be viewed at a distance of $8 \times 85 = 680$ mm, or approximately 27 inches, to be seen in proper perspective. If this rule is not followed, an observer's perception of depth in the image is either expanded or contracted. The apparent depth of a scene is larger when the actual viewing distance is greater than $E \times F$ and smaller when the viewing distance is less than $E \times F$ (Figure 10.2). In spite of these scientific facts, photographic evidence obtained with wide-angle or telephoto lenses may be the subject of intensive questioning by the opposing attorney. Answering these questions may require the testimony of an optical or photographic expert. To avoid this potential problem, it is usually advisable to use a 50-mm fixed-focal-length lens whenever practical as a primary lens or as a backup for other lenses, particularly in those applications where the interpretation of distance may become important.

Another inherent disadvantage of the zoom lens is the difficulty in knowing the exact focal length used in recording a scene. If the focal-length settings on the zoom lens are engraved on the lens barrel, the value actually selected should be recorded in field notes. Unfortunately, most zoom lenses do not have precise focal-length gradations, and focal-length settings are easily overlooked in jotting down field data. These omissions can gravely affect the validity of an expert's testimony when photographic evidence is involved. One solution to this potential problem is to zoom in and out until the desired field of view is obtained, then widen the view as required to set the lens at a specific focal-length setting. This focal length setting along with the relative aperture, shutter speed, film type, and lens serial number should be recorded in the field notes. Reasons for recording all of these data are covered later in the chapter.

A third lens that a forensic expert may often find useful is the macro lens for closeup photography. Conventional camera lenses are optically designed for optimum quality at large object-to-image distances. The focus of these lenses can be adjusted for closeup photography through the use of accessory closeup lenses, extension tubes, or bellows. This

a

b

Figure 10.2 Photographs of a fire hydrant taken with a (a) 20-mm lens, (b) 40-mm lens, and (c) 135-mm lens.

c

ability to extend the usefulness of the lens is achieved at the expense of image sharpness, particularly at the edges and corners of the field. A macro lens is designed specifically for closeup work and does not suffer from this deficiency when applied normally throughout its normal focusing range.

10.3.3 Additional Accessory Items

Other photographic accessories useful to the forensic expert are listed in Table 10.2. The items are shown in approximate order of general importance. Stable tripod, cable release, electronic flash unit, tape measure, and appropriate in-scene reference scales are absolute requirements. The other items listed may or may not be important, depending on the expert's field of specialty.

Table 10.2   Useful Photographic Accessories

50-mm lens (f/1.4 or f/1.8)	Color reference chart
Zoom lens (28–85 or 35–105 mm)	Remote flash unit
Macro lens	White chalk
Lens extension tubes	Felt-tipped pen
Electronic flash unit	Notebook
Tripod	Closeup lenses
Cable release	Lens tissue
Tape measure	Lens cleaning solution
In-scene reference scales	Lens filter
18% gray card	

10.4   MEASUREMENTS FROM PHOTOGRAPHS

10.4.1   Reference Scales

Among the most important nonphotographic accessories in the forensic expert's photographic kit is a selection of in-scene reference scales. The eventual need for making measurements from photographs may not be apparent at the time the exposure is made. Later, when the physical evidence is gone and the photograph is the only remaining source of spatial information regarding a critical measurement that was overlooked or possibly recorded in error, the need becomes paramount. Typical photographic measurements desired are given in Table 10.3.

Inclusion of a linear reference scale in the camera's field of view does not constitute a *true* metric photograph unless the film and object planes are mutually parallel and the scale is in intimate contact with the object to be measured (Figure 10.3). A linear scale in the foreground of an oblique view, for example, does not provide sufficient information to measure distances in other portions of the scene. Metric information on a linear scale is strictly unidimensional; it is limited to one straight line through the scene. When the object and image planes are mutually parallel, image magnification (M) can be obtained by dividing the scale length in the image by the actual length of the scale in the object field. In this unique situation, the magnification (M) is the same everywhere in the object plane containing the scales, so any object distance can easily be determined by dividing the corresponding image distance by M. When the object plane (represented by the floor of a room, a roadway intersection, etc.) and the image plane are not parallel (as in most photographic situations), a two-dimensional scale of reference is required in the object plane to permit measurements to be made from the photographic image. The minimum requirements for a two-dimensional scale are four known noncollinear points in the object plane. A rectangle of known dimensions fits these requirements; the four corners provide the required number of points and a straight line cannot be drawn through more than two of them. A square

Table 10.3 Photographic Measurements

Quantity to be Determined	Examples of Quantities to be Determined	Quantity Required		
		Object Field	Photographic Technique	Photograph
Location of a point (or points) in a plane; distance between two or more points; or the angle between two or more lines in a plane	Size and location of a burn mark on a table surface; lengths and orientations of skid marks on a street surface; or size and distances between holes in a windowpane	Locations of four control points in the plane	None	x and y coordinates of the images of points of unknown location and four control points
Length of an intersecting line normal to a plane	Vertical height of a flag pole above the plane of the ground	Two control lines of known length that intersect the plane at right angles in addition to the above control points in the plane	None	x and y coordinates of images of points of unknown location, four control points, and both ends of two control lines
Location of a point (or points) in space	Height of a stream of water from a fire hose; height above the ground of a nonvertical pole; or height above the floor of a hanging light fixture	Locations of nine non-coplanar control points (six are mandatory, the other three provide desirable redundancy)	Two photographs of the same scene taken from different vantage points (at the same instant of time for moving objects)	x and y coordinates of the images of points of unknown location and nine control points in both photographs (six control points are mandatory)

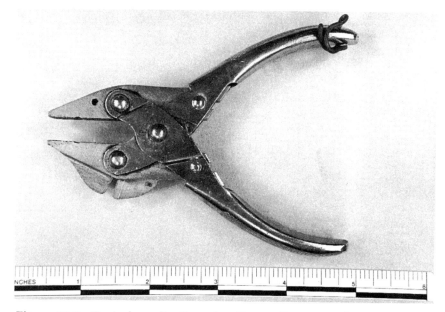

Figure 10.3 Typical application of a linear distance scale in a closeup photograph.

template is commonly used as the basis for the perspective grid technique of photogrammetric analysis to be discussed later. A cross of the type shown in Figure 10.4 also fits the four-point, noncollinear requirement. The two legs of the cross are mutually perpendicular and bisecting. The four extreme reference points on the scale are exactly 2 feet from the center of the cross and $2\sqrt{2}$ feet from each other. The advantage of this type of two-dimensional scale is that it can be folded up for easy carrying. Another form of scale that meets the above requirements is a set of three circular disks placed in the object plane. Techniques of applying these various scales are described later.

10.4.2   Macro Photogrammetric Scale

The L-shaped scale shown in Figure 10.5 is a useful two-dimensional scale for small-object photography. Grid lines in the photograph can be constructed by extending the 1-centimeter gradations on both legs of the scale to correct for oblique camera angles. The position of any point can be measured relative to the closest grid lines by interpolating between them. Distances expanded or contracted by perspective can be measured quite accurately using this gridding technique (Hyzer and Krauss 1988).

Figure 10.4 The author holds a linear scale graduated in hundredths of a foot. A folding, 4-foot, two-dimensional scale is shown in the foreground.

Figure 10.5 Photograph of gouge marks in wood before (left) and after (right) construction of grid lines using the L-shaped scale as described in text.

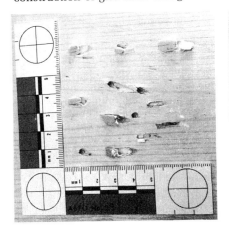 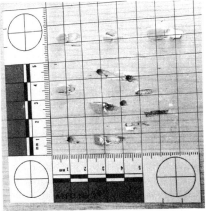

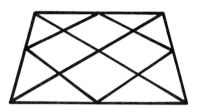

Figure 10.6 Square perspective grid template (left) is shown as it appears in an oblique photograph (right).

10.4.3   Perspective Grid Technique

The perspective grid method consists of photographing a square template of known size located on the same flat surface that is to be mapped. Viewed or photographed from an oblique angle, the square template takes on a wedge shape. Based on the laws of perspective, a complete grid pattern can be constructed on the photograph using the image of the template as the master "square" from which the others are generated. The locations of objects of interest in the plane of the grid are determined by interpolating their positions within a given grid element. A square perspective grid template is shown in Figure 10.6.

Square templates 2 feet in size are commonly employed to map traffic accident scenes (Hyzer 1982). Smaller templates have application in mapping smaller areas. Gridding is usually accurate to distances into the scene not exceeding 10 to 12 times the dimensions of the template. For example, a 2-foot square template can be used to map an area that does not extend more than 20 to 24 feet. Perspective grid templates are available commercially.[1]

The accepted technique in using the template to map scenes larger than 10 to 12 times the template size is to start with a baseline of chalk, paint, or string on the floor or ground to serve as the zero reference line from which all measurements are made (Baker 1985). The template is placed with either its near or far edge along this baseline. A tape measure or string is stretched from the baseline through the center of the template to a distance of at least 10 template widths into the scene. A reference target is then placed on this center line exactly 10 template widths from the

---

baseline. For the case of a 2 × 2-foot template, the reference target would be placed 20 feet from the baseline (Figure 10.7).

The camera position is then carefully chosen to include the template in the foreground center with the template's near edge parallel to the camera's frame line (Figure 10.8). The center line extending through the center of the template to the reference target should bisect the field seen through the camera's viewfinder. If the scene of interest is larger than that covered in the first photograph, the template is moved to the position of the reference target, which becomes the second baseline, and the procedure is repeated.

The gridding of photographs that include square templates is discussed in detail by several writers (Hyzer 1985a; Baker 1985; Whitnall and Millen-Playter 1986).

10.5 RECONSTRUCTION METHODS

10.5.1 Two-Dimensional Reconstruction Methods

There are three approaches to two-dimensional reconstruction from a single image containing an appropriate scale or four known points of reference: (1) algebraic, (2) graphical, and (3) a combination of the two.

The algebraic approach requires the solution of eight simultaneous equations for the determination of eight calibration constants in the following two equations:

$$X = \frac{C_1 + C_2 x + C_3 y}{C_4 x + C_5 y + 1} \tag{10.1}$$

$$Y = \frac{C_6 + C_7 x + C_8 y}{C_4 x + C_5 y + 1} \tag{10.2}$$

X and Y are object-plane coordinates, x and y are corresponding image-plane coordinates, and C_1, C_2, C_3, and so forth are calibration constants that must be determined for each particular combination of photograph and scene. The solution requires four points in the object and image planes for

Figure 10.7 Location of camera, perspective grid template, and reference target in two-dimensional mapping technique described in text.

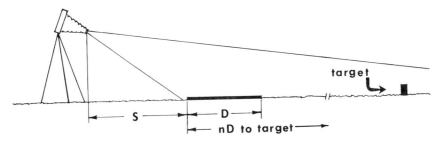

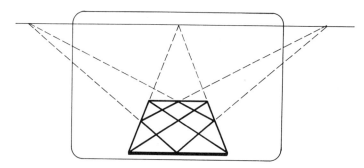

Figure 10.8 Graphical method of locating vanishing line with square perspective grid template using template edges and diagonal lines.

which the X, Y, x, and y values are known. Once the calibration constants have been determined from these four known points, any point in the object plane can be located from the x and y coordinates of that same point in the image. The solution of eight simultaneous equations is a tedious operation to perform manually. Computer software is available and is recommended as the only practical approach to computing the calibration constants.

The graphical method of gridding a photograph that includes a perspective grid is described in detail in several references (Hyzer 1985a, Baker 1985, Whitnall and Millen-Playter 1986). These references are readily available, so the technique will not be repeated here. The pure graphical approach is highly vulnerable to cumulative errors in laying out the transversal grid lines, which prompted the author to develop the combination algebraic and graphical method described next.

The first step in gridding a photograph is the establishment of the location of the *vanishing line*, which is the locus of all vanishing *points* in the image. One point on this line is located graphically by extending lines from the converging edges of the template to their point of intersection. The diagonal lines on the template are also extended to establish two additional points of intersection. A straight line drawn through these three points defines the vanishing line (Figure 10.8). A major source of error with this method arises from the attempt to extend lines along edges of the perspective grid template over distances on the photograph that are large compared with the size of the template. An inaccurately located vanishing line results in corresponding inaccuracies in the location of the transversal grid lines that all run essentially parallel to it.

Accuracies are greatly improved by incorporating a reference target in the scene a known distance away from the square template, as described earlier. It is then possible to calculate the distance between the template and the vanishing line in the photographic image by a method to be described later.

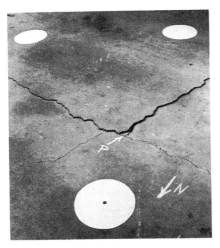

Figure 10.9 Photograph of cracked concrete before (left) and after (right) construction of 1-foot grid lines from three 12-inch diameter circular templates.

Another way to improve the accuracy is to include three circular disks in the scene instead of a square template (Figure 10.9). The author first suggested the circular scale of reference in 1981 as an improvement over scales that were common at that time. Three circular scales of identical size are required in the object plane to reconstruct it from its oblique photographic image. The three circular disks of known size are best arranged as shown in Figures 10.9 and 10.10, with one disk in the foreground center and the other two in the left and right background to

Figure 10.10 Use of circular templates in defining the vanishing line.

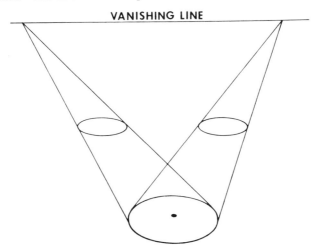

VANISHING LINE

bracket the area of interest. It is not necessary to know the distances between them. This nonrigorous triangular arrangement permits the vanishing line to be determined quite accurately by drawing lines tangent to the three circles as shown. Then two orthogonal lines are drawn tangent to the disk in center foreground from point k in Figure 10.11.

Point k on the vanishing line is defined as the origin of a line perpendicular to the vanishing line that passes through the optical center of the image. Next, two transversal lines are drawn parallel to the vanishing line and tangent to the edges of the disk to complete the "square" m–n–p–q surrounding it. This constructed image of a square template is functionally the same as its optical counterpart, except that the location and orientation of the vanishing line are more accurate. A further important advantage of this construction is that both transversal edges m–n and p–q of the resulting "square" m–n–p–q are always exactly parallel to the

Figure 10.11 Construction of a grid "square" from the image of a circular template.

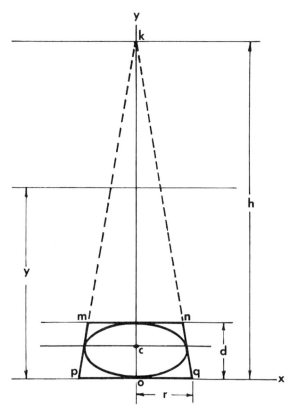

vanishing line (Hyzer 1986a). In other words, the vanishing point for the transversal grid lines is at infinity.

Seldom in practice is it possible to photograph a square template precisely on center line so its transversal edges are strictly parallel to the vanishing line. The result is two vanishing points: one for the transversal grid lines and one for the orthogonal grid lines. This complicates the grid construction if both vanishing points are recognized, or introduces errors if the transversal vanishing point is assumed to be at infinity. The y coordinates of the transversal grid lines are computed from the following equations:

$$y = h - \frac{Sh}{S + nD} \tag{10.3}$$

$$Y = nD \tag{10.4}$$

where

$$S = \frac{nDg - nDd}{nd - g} \tag{10.5}$$

and

$$h = \frac{(S + D)d}{D} \tag{10.6}$$

If a reference target is not included in the scene, h must be determined graphically in the photograph by techniques described earlier. The distance S is then computed by means of the equation

$$S = \frac{Dh - Dd}{d} \tag{10.7}$$

The symbols in Equations 10.3 through 10.7 are shown in Figures 10.7, 10.11, and 10.12 and are defined as follows:

S = ground distance between the camera lens and the near edge of the template
D = outside dimensions of the perspective grid template
d = image distance between the near and far edges of the perspective grid template
h = image distance along the center line between the near edge of the template and the vanishing line
g = image distance between the near edge of the template and the reference target
n = number of each transversal grid line, starting with n = 0 for the near edge of the template
Y = ground distance between the near edge of the template and the transverse grid lines
y = image distance corresponding to Y

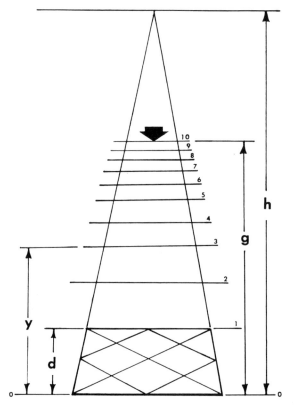

Figure 10.12 Construction of transversal grid lines from the image of a square template.

10.5.2 Two-Dimensional Reconstruction without a Scale

The following two-dimensional reconstruction method is also applicable to single images, but it does not require the inclusion of a scale in the object plane (Hyzer 1963). Instead, (1) the camera must be mounted with the center of its lens a known distance above the object plane, (2) the film and object planes must be mutually perpendicular, and (3) the focal distance (F) of the lens must be known. The first two conditions are easily satisfied if the object plane consists of level ground (the floor of a building, a flat roadway surface, etc.) and the camera is carefully leveled on a tripod a specified distance above it. Reconstruction is performed by superimposing the grid pattern in Figure 10.13 with the image so that the horizon or vanishing line bisects the frame.

The grid is calibrated in units of height of the camera above the ground. If the center of the lens is exactly 36 inches high, all measurements

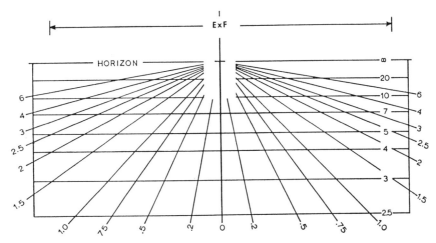

Figure 10.13 Grid pattern for two-dimensional reconstruction using a leveled camera.

may be read directly in yards. If these readings are to be obtained directly from the negative or an enlargement, the magnification (E) of the enlargement times the focal distance (F = the distance between the second nodal point of the lens and the film plane) must equal the distance E × F shown at the top of the chart (Figure 10.13). If the measurements are taken directly from the camera negative or a contact print, the enlargement factor (E) is 1.0. If the camera lens is set at the infinity position, the focal distance (F) becomes equivalent to the focal length of the lens; otherwise it is the focal length plus the bellows extension. The grid is based on the condition that the axis of the camera lens is parallel to the ground and the film plane is perpendicular to the ground. The camera must be carefully leveled to meet this condition.

To use the perspective grid, the picture is enlarged to the proper scale: E × F = scale distance on the grid. A transparent replica of the grid is aligned with the photograph so that the optical center of the photograph coincides with the vanishing point on the grid and the horizontal lines in the photograph are parallel to the vanishing line of the grid. It is assumed here that the camera was properly leveled in all planes. A circular, bubble-type level is convenient for this operation. Distances fore and aft and left and right in the plane of the ground may be measured directly from the grid in terms of camera height above the ground. The grid is applicable to these measurements *only* if the object plane is level and the optic axis is parallel to it. One useful technique is to produce an artificial "horizon" or vanishing line in the object space with reference points located in the background that are the same height above the ground as the camera lens. The vanishing line on the grid is aligned with these points, thereby eliminating the need to find the optical center of the photographic image.

10.5.3 Three-Dimensional Reconstruction from a Single Image

The three-dimensional reconstruction of a scene usually requires a matched pair of images recorded from two different vantage points, but there are specialized exceptions to this general rule. Williamson and Brill (1987) suggest that three-dimensional object-space reconstruction is possible in special cases from single images in which *a priori* information about the geometries of recorded objects is available to the reconstructionist. Architectural structures with straight-line edges and right angles fall into this general category. In general, however, assumptions need to be made that may be questionable. It is far better to photograph the scene from two vantage points and obtain the necessary spatially separated pair of images than to rely on the limited spatial information available in only one image.

The heights of objects that stand normal to flat surfaces can be determined from single images if sufficient three-dimensional information is available for calibration purposes. If the film and object planes are mutually perpendicular, the height of an object can be scaled with dividers by simply transferring its height to the transversal grid line intersecting its base, then measuring the distance between two orthogonal grid lines in the gridded photograph. If the film plane is oblique to the object plane and the focal distance is not known, sufficient information for height measurements is provided by four known points, a perspective grid template or three circular disks in the object plane, *plus* two scales of known heights that intersect the plane at right angles.

First, the locations of points at the base of the object of unknown height and the two scales of known height are determined by one of the two-dimensional methods described earlier. Once all of these base points of intersection with the plane have been established, the unknown height can be calculated relative to the two known heights, using the same vanishing line and laws of perspective previously defined.

10.5.4 Analog Reconstruction

Another technique is useful in obtaining measurements from photographs that depict information that no longer exists at the scene, but do not include sufficient calibration data for performing any of the foregoing reconstruction procedures. This method requires making a copy transparency of the original photograph and returning to the original scene with an SLR camera equipped with a zoom lens (Hyzer 1985b).

The first step is to make a 35-mm black-and-white copy transparency of the print. The film plane should be carefully aligned parallel to the copyboard to minimize distortion in the copy.

The transparency can be used in one of two ways. It either is placed in contact with the focusing screen in a 35-mm SLR camera in which the screen is accessible, or is sandwiched with a piece of frosted plastic film and taped into the camera's focal plane. The first method is preferable. If, for example, a 35-mm SLR is used, the transparency is placed, emulsion side up, under the ground glass. In the latter case, the images are viewed through a hooded magnifier with both the camera back and shutter held open. Accurate replication of the original photographic image is accomplished at the scene by independently varying (1) the camera's longitudinal, lateral, and vertical positions, (2) the camera's orientation in azimuth and elevation, and (3) lens focal length until the real-time image exactly matches the photographic image. Variation of focal length is achieved by means of a zoom lens.

The task of matching the two images is simplified by systematically adjusting each of these three variables in the proper order. First, examine the original photograph carefully. Look for some image element in the foreground that lines up with another recognizable feature in the background. A signpost, for example, may fall directly between two distinctive pine trees on the distant horizon. Locate these same elements in the actual scene. It can then be assumed that the original camera position was located somewhere along an imaginary line extending longitudinally from these two scene elements. Three variables remain to be determined: (1) lens focal length, (2) camera azimuth and elevation, and (3) camera distance along and height above the longitudinal line running into the depth of the scene.

Attention should now be directed to distant objects that can be distinguished in both images. The farther away they are, the better. Adjust the focal length and camera orientation to superimpose these images. Neglect foreground features at this stage because they are likely to be either too close or too far away from the camera to obtain a match.

Now the camera should be moved longitudinally, either backward or forward, until all features in both images begin to match. This is a trial-and-error process that includes varying camera heights above the ground. Slight adjustments in focal length may be required as well to obtain a close match. It is important to keep in mind that zooming the lens varies the magnification of *all* objects in the camera's field of view equally. For example, if the focal length is changed from 50 to 100 mm, the images of a foreground object and a distant object both double in size. A change in camera distance into the depth of the scene has greater effect on the size of images of near objects than on those far away. For example, if two scene elements are 20 and 200 feet away and the camera is moved 10 feet into the scene, the image of the near object will double in size, but the image of the distant object will increase by a factor of only 200/190, or 1.05×. This is why focal length is varied first, before camera distance.

When a close match between images has been accomplished, and the camera has been securely mounted on a tripod, the scene can be rephotographed with an appropriate scale in the field of view. With the help of an assistant, the positions of artifacts appearing in the original photograph can be traced on the ground, walls, and so forth, in the actual scene.

The analog approach to scene reconstruction offers two important advantages: (1) it is a relatively easy technique to explain to a jury or other laypersons through the use of overlapping composite enlargements that demonstrate the superimposition process, and (2) it is a method that most photographers are equipped to handle.

Another approach, which also involves returning to the original scene, requires that the scene be rephotographed in stereo with a calibrated camera (Bailey 1986). Suitable cameras are available from Rollei Fototechnic[2] and Victor Hasselblad, Inc.[3] These cameras are equipped with (1) a lens that has been calibrated for distortion and (2) a reseau or measuring graticule built into the focal plane to establish a precise pattern of reference marks on the film. The stereo pair of images is made by simply stepping sideways a pace or two between exposures with the same camera. Both stereo views must include at least three or four of the same landmarks visible in the original single image, for calibration purposes. Of course, the critical information visible in the first photograph is now gone; otherwise it could be measured directly. Spatial data from these three images are then analyzed by means of BINGO, a computer program originally developed at the University of Hannover and modified by Wild Heerbrugg, Ltd., Heerbrugg, Switzerland. This program digitally corrects the original image by comparing it with the calibrated stereo views, and computes the measurements of required points within the image.

10.5.5   Stereophotogrammetry

Three-dimensional object-space reconstruction from stereo imaging requires a matched pair of photographs exposed either simultaneously or sequentially from two different vantage points. Usually the two photographs are coplanar and recorded at a known distance of separation using matched camera optics. This seems to imply a highly specialized stereoscopic camera, but fortunately this is not necessarily the case. A conventional camera meets these specifications if the camera is moved a known distance between exposures (Hyzer 1986b). One simple way to accomplish this is to translate the camera horizontally. Another is to raise and lower the camera vertically using the rack-and-pinion driven center post on the camera tripod.

Ideally, the shift should be between one fourth and one third of the distance to the principal object in the scene to provide sufficient

---

[2] Rollei Fototechnic, Braunschweig, West Germany, is represented in the United States by Terra Metric, Inc., 3470 Mount Diablo Blvd., Lafayette, CA 94549-3966.
[3] Victor Hasselblad, Inc. is located in the United States at 10 Madison Road, Fairfield, NJ 07006.

parallax between the images. In practice, this requires a wide-angle lens and is not always achievable.

As the same camera makes both exposures, the optics are perfectly matched in the two photographs. Care must be taken, however, to ensure that objects in the scene do not move between exposures.

If a scale of reference is not included in the scene, the X, Y, and Z coordinates of any point in the object space are determined from the following quantities: (1) the x and y coordinates of the corresponding image points in both photographs, (2) the translation distance between the two exposures, (3) the focal length of the lens, and (4) the object distance at which the lens is focused. X and Y are the horizontal and vertical distances between the object point in question and the extension of the optic axis through the object space. Z is the distance along the optic axis of the camera lens. The image distances x and y are horizontal and vertical distances measured in the photographs.

If a suitable set of calibrated distances are included in the scene, it is not necessary to know the lens focal distance nor the distance that the camera is translated between exposures. Using an operation known as *direct linear* transformation, a minimum of nine known points are required in both views, preferably distributed throughout the height, width, and depth of the scene, with no more than four of these points falling in the same plane and no more than two of them falling along a straight line. Ideally, 10 or more calibration points should be included in the scene to provide some redundancy as a basis for determining accuracy and maximizing confidence in the data (Gillen 1986).

Copious points of reference are built into most architectural subjects where the sizes and locations of doors, windows, stairways, and so forth, are known or can be easily measured. Walls or floors constructed of brick, tile, or concrete masonry units of known dimensions provide natural grid patterns that are very useful as available two-dimensional distance scales, which become three-dimensional at the corners where two brick walls intersect or a concrete block wall meets a tiled floor. Measurements should be taken of these features and recorded in the event that other distance measurements may be required from the photographs sometime in the future.

It is not absolutely necessary to shift the camera precisely parallel to the film plane between successive stereo exposures. Simply sidestepping a step or two to the left and then again to the right between exposures is good standard practice in recording any scene that may later require three-dimensional reconstruction from the photographic images. This practice results in three images from three different vantage points that are useful not only in three-dimensional reconstruction of the scene, but also in producing slightly different views of the same scene from which the best single view can be selected.

Analytical systems capable of generating three-dimensional maps from a pair of photographs made with conventional small-format cameras are manufactured by Rollei Fototechnic, American Measuring Instruments, Inc.,[4] and Adam Technology.[5] Photogrammetric services are also offered by private contractors equipped with one of these systems.

10.6 DARKROOM PROCEDURES

There are several advantages to be gained from developing and printing one's own films. One of the most obvious is maintaining an unbroken chain of custody. There are other advantages to those skilled in darkroom techniques. Uniform high quality and fast turnaround time are reasons that most professionals give for processing their own color and black-and-white films.

Forensic experts experienced in darkroom procedures are also able to apply specialized techniques and correctional procedures to optimize their photographic results. Chemical intensifiers and special developers can be used effectively to increase the sensitivities of films beyond their rated values, or to improve image contrast. Two separate images can be superimposed in the same print for comparative analysis and enlargements can be made to any desired scale for measurement purposes. These and other special procedures are not readily available to those who depend on outside processing laboratories.

10.7 RECORD KEEPING

The need for proper recognition, collection, and preservation of physical evidence is critical to any forensic investigation. A good source of information on that and related topics, including a section on crime-scene photography, is the *Criminal Investigation and Physical Evidence Handbook* (Wisconsin Department of Justice 1981).

Important photographic data that should be recorded in the field notes are listed in Table 10.4. There are no hard and fast rules for record keeping except that it is usually better to have more recorded information than not enough. The use of automatic cameras makes it difficult to record specific exposure information. Under these conditions, it is usually sufficient to merely note that the camera has been set on automatic exposure and/or focus; otherwise, specific camera settings should be recorded in the field notes. These data are usually far less important with video cameras because it is possible to preview the image in the camera's finder, and it is very unlikely that the video records will ever be subjected to metric analysis as might be the case with film cameras. A time–date code generator which electronically produces a numerical display on the tape

---

[4] American Measuring Instruments, Inc. is a technology affiliate of H. Deli Foster and Associates, 2400 Freedom, San Antonio, TX 78217.

[5] Adam Technology, Bently, Western Australia 6102, is represented in the United States by E. Coyote Enterprises, Inc., Route 4, Mineral Wells, TX 76067.

along with a digital watch timer is useful in authenticating videotape records. These data verify that the tape has not been edited and is being played back at normal speed. Inclusion of a color patch checker or a gray card at the beginning of each scene recorded under different lighting conditions is also recommended as verification of true color balance.

The recorded data listed in Table 10.4 are intended as examples that may or may not apply to specific photographic situations. This tabulation is intended merely as a checklist of notebook entries that might be required in specific instances.

Film identification is often important in maintaining chain of custody and in the correlation of photographs with notebook records. One highly recommended technique is to reserve the first frame on each roll of film for recording a reference number. Some practitioners write the number on a note pad with a broad felt-tipped pen and then photograph it at close range. A further point of identification is achieved if the photographer's name and address are included along with the current date.

I personally prefer to incorporate this time information into the reference number by using the year, month, day, hour, and minute as an identification number. This always results in a unique number that is easily traceable to a specific time. For example, a film loaded into the camera at 1:30 PM on March 15, 1987 would be numbered *8703151330*. I write the number on a white card or sheet of paper with a felt-tipped marker. A Kodak 18% gray card, or in some cases a graduated gray scale, is placed alongside the page with the identification number. Also included in

Table 10.4 Field Note Data

Camera name and serial number
Lens name and serial number
Film type and identifying number
Lens f/number
Lens focus setting
Lens accessories employed (filters, closeup lenses, extension tubes, bellows, etc.)
Film ISO value used in computing exposure
Illuminating source (sunlight, overcast skylight, electronic flash, incandescent light, combination of sources, etc.)
Camera shutter speed
Description of spatial reference scale
Distance measurements within the picture area
Pertinent environmental conditions
Date and time of day
Camera orientation (azimuth, elevation, etc.)
Film processing information (name of processing lab and receipt number) For those who process their own films, film developer type, development time, and temperature should be noted along with pertinent information regarding film printing techniques.
Persons present when photographs were made
Any other pertinent information that might be useful later in verifying photographs

the field of view is a business card with my name and address. These data are photographed under exposure conditions prevailing at the actual scene. The gray card and scale are useful to the processing lab in producing color reproductions of correct color balance. The film reference number, along with pertinent photographic information, is recorded in my field notes.

The same lighting conditions may not prevail in every film exposure. A typical investigation might require some photographs to be exposed under existing light conditions and others to be exposed using an electronic flash unit. Under varying conditions, it may be useful to photograph the gray card again on the same roll, noting the new lighting conditions. Care must be taken in orienting the gray card or scale relative to the light source to avoid direct reflections from light striking the surface. Ideally, the gray card should be oriented at a 45° angle to the axis of the illuminating source.

I usually request that the film be returned from the processor in one continuous strip thereby maintaining the integrity of the film numbering system. If the film is later cut into smaller strips, I number each of these individual strips using an indelible felt-tipped marking pen. The same number used to identify the roll is printed just outside of the perforations on each strip cut from the master roll.

Each frame on a 35-mm roll is identified with a number exposed at the edge of the film by the film manufacturer. An individual frame is easily identified by the film identification number followed by a hyphen and the frame identification number, for example, 8703151330-25. This number identifies the picture as frame number 25 from roll number 8703151330. This same number may be used to identify prints and slides produced from that frame of the film.

All receipts for film processing, reprints, enlargements, and so on, should be pasted in the notebook for safe keeping. I also paste in the ends of the film cartons showing the film type and emulsion number—or include them in the reference photograph on the first frame of the film—as a standard record-keeping procedure.

Obtaining written permission to photograph individuals or to photograph on private property is also important; otherwise, the photographs may not be admissible in court. I keep these records as an integral part of my field notes.

10.8 ROLE OF THE IMAGING EXPERT

Experts in disciplines outside of imaging science are often reluctant to apply any of the reconstruction methods cited in the previous sections to photographs taken by others, and even in some cases to their own photographs. Their reasoning is valid for several reasons. The intricacies of defending the scientific validity of these techniques under cross-examination require the expertise of a specialist in imaging science who is qualified to testify in matters pertaining to light, optics, physics and chemistry of imaging, photogrammetry, and the psychophysics of human visual perception in the interpretation of images. For purposes of a trial, a photograph is viewed as a graphic portrayal of oral testimony and is admissible only when a witness has testified that it is a correct and relevant fact personally observed by the witness (Fletcher 1986).

Imaging technology has not reached the stage of development where it can be stated, in the strictest sense, that a photograph, or any image for that matter, is a true and accurate representation of any recorded object or scene. Photography, which is the most faithful of any imaging medium in its reproduction qualities, is a two-dimensional system, whereas most objects and scenes are three-dimensional. Stereophotography is a step forward toward reality, but the intricacies in viewing images in three dimensions discourage its use in courtroom situations except in rare instances where the visual communication of highly specialized information is required.

There are other deficiencies of photography in accurately portraying any object or scene. These include limited tonal range and imperfect color reproduction. The luminance ranges of most scenes far exceed the dynamic response ranges of photographic films and prints (Hyzer 1987b). The color reproduction qualities of various color films vary widely, are highly dependent on the color quality of illuminating sources, and are sensitive to the controls exercised in processing. The ability of any imaging system to accurately reproduce a range of colors is approximate at best (Hyzer 1987c).

In spite of these limitations, any witness, expert or not, who is familiar with the facts represented or the scene photographed can testify as to the photograph's correct and accurate representation of the facts. Neither the photographer's nor the expert's testimony is required. However, when measurements and reconstructions are performed on photographs, it becomes quite a different matter. The foundation for admissibility is the ability to demonstrate that a reliable scientific process was correctly applied to obtain the results offered in evidence. For these reasons, imaging scientists or photogrammetrists are usually called on to perform analyses on existing photographs.

It has been said that the "camera never lies." Perhaps this is so under the critical eye of a knowledgeable imaging expert, but it cannot be denied that photographs can be manipulated in many ways to distort the truth and mislead unwary viewers. Photographs submitted as evidence should not be accepted as true and accurate representations of anything without subjecting them to critical examination.

The legitimate manipulation of photographic imagery is as old as photography itself. The manual retouching of negatives to cover up pinholes, remove scratches, and obscure facial blemishes in portraits has developed into an art in itself over the past century and a half. Special-effects crews in Hollywood have also developed a myriad of image manipulation methods for artistic purposes in the motion picture industry. Consequently, it is not unusual for an attorney to ask whether a photograph produced by the opposing side has been "doctored" in any way. The term *doctoring* is used here because it is common vernacular in legal circles. It implies that the image has been manipulated in some way to emphasize certain details or subdue others. Extreme cases of image manipulation might include eliminating certain details altogether or adding others that were not present in the first place. All of these things are possible with modern digital image processing methods. Elements within an image can be fabricated, enhanced, distorted, shifted, cloned, erased, and/or transferred to another image with a precision that almost defies detection even under the closest scientific scrutiny. Photographs that are suspect of such treatments should be examined by a qualified imaging examiner.

There are more subtle forms of image manipulation that might be better described as *image misrepresentations*. the various effects of perspective have been mentioned earlier and are illustrated in Figure 10.2. Spatial distortions can be used in subtle ways to selectively influence the viewer. Image contrast is also subject to distortion that can be used effectively to misrepresent the true character of an object or scene.

Although some of these image manipulation techniques are proprietary, many have been described in the literature so they can be readily adapted to other purposes by anyone knowledgeable in photography. My purpose here is not to explain how these manipulations might be accomplished; rather it is to alert the reader that images can be manipulated in numerous ways and to varying degrees by a variety of techniques that are readily accessible to anyone who might be interested in applying them for illegitimate or fraudulent purposes.

The gross addition or removal of image elements usually results in a change in the statistical distribution of the grains forming the image at the boundaries of the altered area. This may also be accompanied by an abnormal density gradient across the border separating the altered and unaltered parts of the image. Manipulation that consists of adding new elements to an image is often revealed in the shadows and highlights. An example might be the fraudulent addition of a fire extinguisher to an interior photograph of a building that was later ravaged by fire for the purposes of demonstrating that local fire codes were not violated. To accomplish this, the images of the room and the fire extinguisher could have been taken from two separate camera exposures and later combined by optical or electronic imaging techniques. Unless almost identical lighting was used in photographing both components, highlights and shadows often give away the fakery. A typical example of an image element that might be removed or obscured for ulterior purposes is a visible crack in the supporting piers of a bridge that later collapsed.

Photographs are sometimes submitted as evidence of conditions that existed at the time that the photograph was exposed. A suspect in a crime, for example, may use a photograph to prove that he or she was somewhere else when the crime was committed. A photograph might also be submitted to show the conditions at a construction site before an accident occurred. In these hypothetical examples, the dates when the photographs were taken would be supported by verbal testimony. It is possible in some cases to check the validity of this testimony by examining the photographs themselves. The position of the sun, for example, revealed by shadows, highlights, and reflections is often a key to the time of day and year the exposure was made. The position of shadows in a photograph of a building might indicate that the photograph was actually made early in the afternoon either in mid-March or late September, contradicting verbal testimony that the picture was taken in the afternoon of June 17th. Date codes on films and prints are also very useful in this respect. The date code on a Polaroid print, for example, might reveal that it was actually manufactured after it was reported to have been exposed, thereby disproving verbal testimony.

It may be tempting in legal cases involving the limitations of human vision to use photography as a means of demonstrating what a person could or could not see under certain lighting conditions, especially low-light-level situations where an accident occurred. The attorney may ask for a photograph of the accident scene that is representative of visibility levels that existed at the time of the accident. Research has shown that there is a wide disparity between the physics of photography and the psychophysics of human perception (Hyzer 1983a, 1987a; Lieskovsky 1985). The production of demonstrative evidence of this type should be left to an imaging expert who is knowledgeable in both photography and human visual perception. It is questionable whether a true and accurate photographic representation of a low-light-level scene is achievable, except in very rare situations.

10.9 GUIDELINES FOR EFFECTIVE FORENSIC PHOTOGRAPHY

The task of the forensic photographer is to produce images that (1) are the result of the photographer's best efforts in depicting reality and (2) include sufficient information to permit complete and accurate analyses. The following twelve guidelines are offered to assist in accomplishing these objectives:

1. Try to anticipate how each photograph will eventually be used; then choose the best camera position and photographic technique accordingly. (See Table 10.2 for detailed suggestions on camera accessories.)
2. Select the best film for the situation at hand, keeping future requirements in mind. Most forensic photographers prefer 35-mm negative films (black-and-white and color) as opposed to direct reversal films in those cases where enlarged prints may be needed.

3. Critically preview each photographic situation before pressing the camera button. Examine the scene carefully for details that might be difficult to explain later. (A clock or calendar in the scene may *not* indicate the actual time of photography.)
4. Identify each roll of film with an alphanumeric code that coincides with written notes and is traceable throughout its chain of custody.
5. Bracket exposures ±1, and in some cases, ±2 stops.
6. Photograph important scenes or objects from several different vantage points.
7. Include a proper scale and/or make appropriate measurements in all photographs that may be used for quantitative purposes at some later date. (If in doubt, it is better to include a scale than to leave one out.)
8. Scenes including a scale should also be photographed again without a scale in place, to avoid later questions regarding critical evidence that the scale itself might have obscured.
9. Record all pertinent photographic data. (See Table 10.4 for detailed recommendations.)
10. Maintain a file of receipts for film processing, printing, shipping, and other items.
11. Keep all original camera negatives and transparencies on file for a reasonable time. A period of 10 years is good practice.
12. If camera originals are given to an attorney or another expert, obtain a written receipt to maintain an unbroken chain of custody.

References

Bailey, J. 1986. Photometrics and traffic accident investigation. *Technical Photography* 18, No. 8 (August).

Baker, J.S. 1985. *Traffic Accident Investigation Manual*. Evanston, IL. The Traffic Institute, Northwestern University.

Dragel, D. T. 1965. Science and the law. *Analytical Chemistry* 37, No. 3: 27A–31A.

Eastman Kodak 7-71. *Crime Movies—An Aid to Justice*, Publication M-20.

Eastman Kodak. 10-72. *Photographic Surveillance Techniques for Law Enforcement Agencies*, Publication M-8.

Fletcher, J. E. 1986. Basic requirements of photographs to become admissible evidence. *Technical Photography* 18, No. 11 (November).

Gillen, L. G. 1986. Photogrammetric mapping of vehicle deformations. In *SAE Technical Series*, Publication No. 861421. Warrendale, PA: Society of Automotive Engineers.

Hyzer, W. G. 1963. How to Measure with a Camera. *PMI: Photo Methods for Industry*. 6, No. 1 (January).

_____ 1975. Your day in court could be close at hand. *Photomethods* 18 No. 8 (August).

_____ 1982. Perspective grid photography. *Photomethods* 25, No. 7 (July).

_____ 1983a. The eyes have it. *Photomethods*, 26, No. 5 (May): 5, 67–70.

_____ 1983b. Interpreting human nighttime vision. *Photomethods* 26, No. 6 (June): 16–18.

_____ 1985a. Perspective grid photography. *Imaging Technology in Research and Development* (July).

_____ 1985b. The perspective grid and analog reconstruction. *Photomethods* 28, No. 11 (November).

_____ 1986a. The use of circular scales in the perspective grid technique of making photographic measurements. *Journal of Forensic Sciences* 31, No. 1 (January): 20–26.

_____ 1986b. Stereophotogrammetry.*Photomethods* 29, No. 1 (January): 6–7.

_____ 1987a. Modeling the magic of vision. *Photomethods* 29, No. 3 (March): 8–9.

_____ 1987b. Contrast degradation in image projection. *Photomethods* 29, No. 5 (May): 8–9.

_____ 1987c. Accurate color renditions in scene reproductions. *Photomethods* 29, No. 6 (June): 12–13.

Hyzer, W. G., and T. C. Krauss. 1988. The bite mark standard reference scale—ABFO No. 2. *Journal of Forensic Sciences* 33, No. 3 (March).

Jacobson, R. A. 1973. The prestige way to moonlight: Be an expert witness. *Machine Design* (November 13): 132–136.

Lieskovsky, V. 1985. Visual observations under marginal illumination and their forensic evaluation. *Forensic Science International* 29: 109–127.

Whitnall, J., and K. Millen-Playter. 1986. Recording accident data with a perspective grid. In *Photographs and Maps Go to Court*, pp. 35–53. Falls Church, VA: American Society of Photogrammetry and Remote Sensing.

Willimason, J. R., and M. H. Brill. 1987. Three-dimensional reconstruction from two-point perspective imagery. *Photogrammetric Engineering and Remote Sensing* 53, No. 3 (March): 331–335.

Wisconsin Department of Justice. 1981. *Criminal Investigation and Physical Evidence Handbook*. Madison, WI: Crime Laboratory.

11. The Report

M.D. MORRIS, P.E.

11.1 BASIC NOTIONS

You, the forensic engineer, are paid well for what you know, not for what you do. The only tangible item your clients get for their money is your *report*. It is also the only thing that puts your effort into the record. This three-dimensional "package" of hard news, good news, or bad news is the vehicle that carries your observations, your conclusions, your recommendations—your experience. It speaks as your omnipresence simultaneously to every person concerned with the matter. It must be good!

No longer can you afford to regard your report document as a postscript, an additional item at the end of a completed project. Nor can you deal with it in the archaic academically structured tradition: "Ah, I have to write a report. I confect an outline from the extent of my knowledge, plug-in source material, and write." Wrong! Then your writing would consist of an introduction, a body, and a conclusion; all written in the passive voice, third person impersonal. Wrong again!

That method worked before a better, more natural approach emerged, just as the horse was the only power for land transportation from mankind's beginning until as recently as the early nineteenth century. Today's software must keep pace with its hardware. If your report is sparse, your client will come back to you for more information, but not for more assignments. If your document is voluminously overburdened with circumlocution, your message vanishes in the weeds of words, and your professional value disappears along with it.

Start by thinking of the report and the investigation as the same system, instead of envisioning the report as a caboose ending a freight train. Whether it be investigation of a failure, a crime, a labor dispute, or a preconstruction site; or a travel account, a proposal, a master's thesis, a research and development project; or even the running of a small business, the activity is nothing more than the data-gathering stage for a subsequent document. Conversely, the document is nothing more than a record of the preceding activity.

Accepting as a primary premise the notion that the activity and the document are both ends of the same stick, it follows that the project plan is also the report outline. Both germinate and grow together; the report develops as the activity evolves.

By applying this basic idea, and its resultant method, you organize your investigation, your analysis, and your report, all at one time. You will find your project perception more accessible and logical, and thus easier. With this perspicacity, you will break free of the gravitation that limits both your purview, and its consequent, constrained conveyance to your client.

Unlike the traditional method that functions only if you know your subject well, this procedure will work for you every time, without fail, whether you know nothing or all there is to know about a subject. The process is a naturally flowing progression of ideas that I shall list first, and then discuss in detail.

First you have to determine *why* you are writing: *What* do you want your document to do for the receiver? Then you consider the receiver: *Who* will read it all, and who will read selected parts? Now you can decide what to tell this audience.

To do this properly (and not just elaborate on what you know,) you have to stand in the receivers' shoes for the moment. Say, "I'm not writing this, I'm reading it. What do I want it to tell me?" Maybe responding completely might be beyond your knowledge perimeter. If so, then you must expand your horizons. Generate random questions that readers at all levels might ask. Categorize those questions into groups of similar nature. Naming the groups will yield the main sections (chapters) of your report; these are principal matters. These questions grouped under section headings constitute your preliminary (provisional) outline.

At this point you need to define your goal. To establish your expected result you write an approximately fifty-word *scope summary* as a target marker to remind you, along the line, where you want to be when you are done.

With a reasonable research plan, a facility for collecting and inventorying your acquired material, the preliminary outline, and the scope summary to keep you on course, you may now conduct your investigation with some positive direction. This will provide you with the answers to the questions. By organizing those answers in the most useful logical, chronological, or sequential arrangement, you can develop a firm outline from the material at hand. This will also indicate clearly where any discrepancies might be.

Now it all comes together. Use your scope summary as an end-result specification, your development chart as a guide, and your firm outline as structure. From your collected source material write your first draft of all interior sections in the active voice, first to second person. Follow this with your conclusions and recommendations. Then write your introduction to an existing piece of work.

Infuse your graphics here. Depending on how you write (word processor, or tape and type, or handwrite and type), rework, edit, and polish in no more than two further drafts. This whole process should yield a finished product that reflects its metamorphosis as one with your investigation and your findings. It will be a monolithic entity that is responsive to the charge in your assignment; penetrates any barrier of receiver resistance; is at the real receiver's knowledge datum and comprehension level; is sufficiently brief; and involves your target audience.

Surely you will have assignments for which a report is not required; however, for your own protection you should document your efforts "for the record" and keep the file for at least the time required by the legal statute of limitations. Or, you may have a major case that needs superb documentation to restore public confidence in the subject. In all outings, you have a tremendous running start by having the entire program plan in your control.

In essence the foregoing tells the whole story, but it is only information (what you ought to do). You really want instruction (how to do it). To facilitate a fuller explanation, let us consider closely one prototype project as a vehicle. The simplicity of this system will enable you to handle the most difficult cases with ease and understanding. Thus, deliberately avoiding the obvious, I chose this prototype over several other simpler situations. It is an obscure project that required a convoluted presentation to overcome any possible political or policy problem, and still achieve a positive solution without casting a villain.

This is a bona fide "from-the-job" report, accomplished about a dozen years ago by Lynn C. Jewell. The display is accurate. Though it has been "sanitized" well to preserve propriety and protect personal rights, for this practical purpose there is verisimilitude.

The scenario involved an organizational flaw rather than a design or material flaw. And, once pinpointed as a "people problem," it required both fact and tact to resolve: A telephone company area repair service had experienced a deluge of breakdowns both in the plant lines and with user installations.

To anticipate top management's eventual displeasure at learning of this, middle management wisely, and quickly, decided to investigate and to submit a report that included a suggested solution. The first guesses were defective replacement materials and sabotage.

Jewell was assigned the task. Instead of a "traditional" probe followed by a "standard" report, she took a more thorough route.

11.2 THINKING AND PLANNING

Thinking and planning are basic to producing a document that transmits an idea, in its entirety, from one mind to another. Every project must have a *logical basis*: clear-cut reason(s) for writing, and a well-defined target readership on all levels. Those two elements then delineate the specific subject area. What you write depends on why you write and to whom you write. Not vice versa. A *development chart* is the graphic display of that information.

11.2.1 Reasons for Writing

To say that you write because if you don't you'll lose your contract, or be fired, or fail is not correct. These are motivations, not reasons for writing. In effect, a reason is a form of interaction between writer and reader. My experience over 22 years has revealed six principal *reasons* for writing nonfiction:

1. *To Inform.* To impart fact or data that increase or enhance a reader's knowledge. To convey items of which a receiver is not already aware.
2. *To Instruct.* To teach a reader how to accomplish a task (to give directions) or to describe a process or occurrence. This is the way to transmit the notion of motion through a sequential series of still statements. A motion picture film strip, for instance, contains no motion. It conveys movement via a series of still frames, each detecting a slight change of position over a very short time interval. (This is exemplified in calculus as dx/dt.)
3. *To Influence.* To steer your readers' thinking in your direction. Do this by logical persuasion, not distortion of fact, misinformation, or lies. An ethical point arises here: someone consciously, without altering facts, could convolute a true result to appear as something else. There is nothing arcane about trying honestly to influence a receiver. Every proposal must convince its reader to "buy" what is offered, or it fails as a bid, thus wasting its writer's effort. Influence is also the way to work your opinion into the text, when none is required by the assignment. Do this by innuendo, or with an unanswered rhetorical question—subtle infiltration.

   An unplanned aspect may be the influence of the situation on the writer. If you work with a subject for a while, you may develop a distaste for it that could easily result in turgid writing. That, in turn, rubs off on the reader; your negative attitude is transmitted by attrition. Conversely, if you gain a good amount of enthusiasm for the subject, your writing will be ebullient and thus have a positive effect on the reader. Be objective and in charge of your writing: No report is better than the integrity of its writer.

4. *To Control.* Control follows from the preceding discussion of influence. Other than via influence, no one enjoys the privilege of controlling those above him or her on the ladder. But control of people or situations below does not grant a license to be imperious. The easiest way past the barrier of receiver resistance is with an amenity. Including "please" in every request, directive, or order begets better, faster, and more efficient results. Most cases of control in a forensic report appear in the conclusions and/or recommendations sections. They should be positive, clear, and strong to establish causes and to avoid recurrences.

5. *To Criticize.* By nature a forensic report may be critical of the people, design, materials, or functions involved even peripherally in a failure. But this must be constructive criticism. Finding fault alone does not prevent the recurrence of a flaw. To be eligible to criticize, you must separate opinion from fact and be able to offer a better suggestion. You must not cite a personal experience unless it contributes directly to the solution. You must not engage in personalities. Character assassination is not your objective; you must not make jests at another's expense. You must not compound error; be absolutely sure. We live in an overly litigious society that includes a substantial number of easily angered people. Give no one the least chance to "hang one on you."

6. *To Record.* Early paragraphs of this chapter detailed the importance of putting your efforts (in every case, howsoever slight) into the record for your security. These matters never seem to stay buried, and occasionally return to haunt you. Do not be too brief in an in-house report. Should the time come to exhume that record, none of the original participants may be available to fill the voids with missing facts. That, of course, is only slightly worse than having to plow through jungles of miscellaneous useless pages until you find the nugget.

Your selection of any one (or as many as all six) reason(s) for writing will *a priori* establish the use to which your receivers will put your document. That, in turn, will determine its longevity, its useful life in the outside world. In house, it is the period set by the statute of limitations, or longer if you write well and have an eye toward history.

11.2.2 The Audience

The receivers are your readers. The signatory of your assignment, or contract, may not necessarily be the person who actually reads your report. That signer could be a public official, a corporate officer, a commanding officer, a dean, a division director, or anyone authorized to act for an organization, who thus becomes the addressee of record for communications aimed at the organization. That person whose name appears on your missive is the *apparent* audience. In some cases the apparent audience may read your report first or last, or may not even know it exists.

In nearly every instance, though, someone, hopefully well qualified, is delegated to read thoroughly, examine closely, ruminate on, and evaluate the content of your report. This is the *real* audience. It could be one or two persons, a committee, or a board. When you write to a *multiple* audience, choose one person as a datum reader, write to that level, and let the others read "up" or "down." You cannot reach or please the entire known world with one all-inclusive paper and still preserve its tight focus without dilution.

The real audience is obligated professionally to be interested in and to read your whole document thoroughly. Again, in most cases there may be other people equally involved in the project, but as specialists or experts with limited text interests: budgetary types seeking solely fiscal facts; attorneys looking for legal loopholes; designers delving for material matters. These people read only the sections and appendices dealing with items of interest to them, and pass on the rest. They constitute your *secondary* audience.

Finally, you have a *subordinated* audience who may read your document later, for reference or academic interest. Your target is the *real* receiver. The apparent, the secondary, and the subordinated readers all read on their own.

Having isolated and defined your readership levels, dig some to try to ascertain vital signs about your real reader, also information about the other readers. Learning their backgrounds and functional abilities helps you set a document comprehension level. Surrounding conditions may inhibit your readers' reaction and response. Know if there are any of these; think of what can help you circumvent these impediments. Finally, finding the real readers' knowledge datum will guide you to a proper starting stratum: too low will make a bored reader quit; too high will open a knowledge gap that creates enough frustration to cause a sudden stop. Envisioning a knowledge datum will tell you reasonably well what your real readers need to know and want to know. Any adult who has ever been engaged in a romance knows there are differences between needs and wants.

Now, knowing why you are writing and to whom, you can lay out your *development chart*. This will help you see where you are and aid you in selecting a working title. Look at the development chart for the Jewell project at the top of the next page.

Lynn Jewell's listed reasons for writing are obvious. She had no call to *instruct* or to *control* anyone. Any *critical* opinions would fall into *influence*. Primarily, she wanted to inform her readers of the cause of the breakdown, then influence them to accept her suggested solution. To avoid

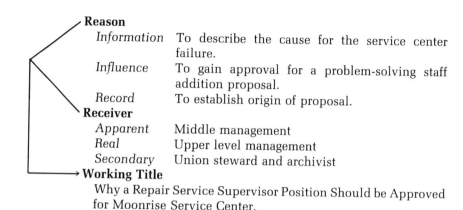

Reason

Information	To describe the cause for the service center failure.
Influence	To gain approval for a problem-solving staff addition proposal.
Record	To establish origin of proposal.

Receiver

Apparent	Middle management
Real	Upper level management
Secondary	Union steward and archivist

Working Title

Why a Repair Service Supervisor Position Should be Approved for Moonrise Service Center.

the remote possibility of someone's taking her work and running with it, she committed her document to the record.

Company protocol specified that top management alone was to make policy decisions. It also stated that top management was to be addressed only through middle management screening, not directly. This meant that Jewell had to write for middle management who would read and forward the report; still, the report had to be written at a level that appealed to the decision makers. Copies to the union stewards and archivist verified her position as originator of the remedy and ensured her credit. Resourcefully, she chose to subrogate the investigation and findings, using them instead as bases for proposing the resolution. Ergo, instead of a "trouble" report with a recommended remedy, she presented management with a proposed organizational change that would preempt future problems surely to arise from the status quo. Her choice of a working title indicates this thrust.

11.2.3   Substance

Now you have begun the process of thinking through and designing *both* your investigative plan and your report of it, simultaneously. You *do* know your discipline and your experience. But you do *not* know what really happened, and what about it your receivers need and want to know. Assume you were in their place. What would you ask, that the report must answer?

To carry through with this notion, generate questions in behalf of all concerned readers. List them as they occur to you, in no special order and with no constraints. Experience shows that 90% of the questions you list originate in the confines of your own knowledge, with a 5% fringe on the right and left.

On the right is the "Ph.D. syndrome" where you become so

steeped in a subject, it becomes yourself. In this case, you cannot back off sufficiently to look at the situation objectively. Every question you originate carries your second thought, "That's too elementary" or "Obviously!" The solution: go to grass roots. Peers of the real readers can provide plenty of questions from the target level. Better, if at all possible, by proximity, propinquity, or personal acquaintance, try to obtain real questions directly from the real readers.

The left-wing 5% "subnovice situation" arises when you know too little about a subject. Not only do you not know what to ask; you do not even know what *kinds* of questions to list. And in this situation you cannot go to grass roots; it would be the blind leading the blind through a jungle. You resolve this by basic research to develop a surface patina of knowledge, not to find facts for the document, but to learn enough to be able to pose intelligent questions. No professionals in their right minds would go to an encyclopedia for valid source material. But for this purpose, you can learn enough to generate good questions that your real research can answer later. At this point there could surface some new hitherto uncovered facet of the entity under scrutiny that might otherwise have escaped unnoticed.

A *preliminary (or provisional) outline* is the product of the question process. List your questions on a lined sheet of paper in the order they come to mind. Do not number the list. Do not preimpose a subtopical order. Do not worry if a question seems repetitive; you will sort that out a bit later. Do not look back. Just list the questions as they come to mind. They may flood at the start. As friction sets in, they slow to a trickle. Stop for awhile and do something else; soon new questions arise. At some reasonable time, stop the process. Too many inquiries could stifle the rest of your time; too few will not solve the problem.

In her prototype project, Jewell generated 41 questions (Figure 11.2.31) that all three receiver levels might have asked. Every project has a life of its own. In one case, the "proper number of questions" might be 25; in another case, 600. The total number is governed by the reason for writing, the real readers, and the project's magnitude. The total is not fixed. You can infuse new questions until the final copy.

For the random list to be of any value, it must be organized or sorted into groups of similar or like entities. Just as you would sort a keg of loose hardware into buckets of bolts, screws, nuts, and washers (but no "etc."), do the same with your question list.

There is neither formula nor algorithm for the number of groups, but there is a practical rule-of-thumb: about 25 questions yield a minimum of three and a maximum of five groups; 75 questions give you five to eight categories. Set out clean sheets of lined paper, one for each anticipated group. Plan for the maximum number of groups each time. Should you have fewer groups than your predicted maximum, you haven't posed your questions over a broad enough area. Do some more. If you have a few more than maximum, don't worry. If there are many more, your focus needs to be narrowed.

How much would proposal cost?
What is organizational structure?
How does it compare with other service centers?
Those with Repair Supervisors?
Those without?
What effect on productivity?
How will cost be returned?
If direct return, in what ways?
If indirect return, in what ways?
How can additional person benefit the district?
How can additional person benefit the company?
What is the present RBA index?
How will overtime be affected?
What is the missed appointment rate?
How will a Repair Supervisor reduce customer takeovers?
What is the percent (%) of customer takeovers?
What is the repeated report rate?
Will personnel development and training increase?
Is there a need for more development and training?
How do the indices and trouble indicators compare
    with other service centers?
What is the coin trouble rate?
What effect on computer usage and analysis time?
How will an additional person benefit the personnel?
How will a Repair Supervisor improve customer relations?
Why can't the present staff adjust to meet the need?
Is there additional cost beyond the outlay of salary
    and overhead?
Adequate space?
Cost of desk and other facilities?
Where is this person to come from?
New hire?
Promotion from ranks?
Lateral transfer?
Should this person be permanent or temporary?
How can this person be used in the future?
Will this person continue to fit into the corporate
    game plan?
Will additional person "deflate" the job of the
    testing foreman?
Who will be responsible for training?
Will additional personnel be necessary?
Will the position be purely supervisory?
Does the present system need restructuring?
What are the alternatives?

Figure 11.2.31 Prototype project: list of questions. What the reader might need (or want) to know. (Courtesy L. C. Jewell.) (Please note that the figures in this chapter are not numbered consistently with the numbering system used in the rest of this book. The system used in this chapter is employed at the insistence of the author, for the purpose of illustrating his preferred method.)

This is a good spot to determine if your project plan will lead you to the solution you seek. It is better to rethink now if you must, rather than backtrack from a wrong choice.

Go down the list; group questions of the same nature in sublists. Some questions may fit into more than one group. List those questions in all the groups they fit.

When you have completed this primary (generic) sorting, you may discard your original list without losing anything. Now review each sublist and remove redundancies. Consolidate, replace, or add new questions if the opportunity arises. After the sublists meet your satisfaction, reread each one. The nature of the material within a sublist will suggest its title. In effect, the name floats to the top of the list.

Jewell sorted her 41 questions into six groups; one of the groups contained only one question. There is nothing wrong with a single question under a heading, if indeed that question is unique. (I recall that my college physics text, by Houseman and Slack, contained a chapter just half a page long.) In Jewell's case, the single question was not unique; she managed, without struggle, to include it in another sublist. This left the five categories shown in Figure 11.2.32. After some additions, deletions, and changes, the net gain/loss in the original number of questions was zero. The collective total, by coincidence, remained at 41.

By reading the questions, the sublist titles surfaced as you see them—Organizational Structure, Cost, Analysis Time, Customer Service, and Advantages—not necessarily in the final order for the report. These generic titles were Jewell's main project subtopics, or principal matters. Also, they were her areas of investigation, all identified and labeled.

A *scope summary* could be prepared now. This item never appears per se in the final report, but it is a multipurpose necessity. It is your aspiration or goal set (a target marker like the pin flag on a golf green). You could even regard it as a short end-result specification, but it is really a memorandum from you to yourself. Write your scope summary on a 3 × 5-inch file card for handiness and durability. In 50 words or fewer, tell yourself where you want to be when you get there. Try to weave in your generic reasons for writing.

If you write your scope summary with care, in almost all cases, you should be able to use it intact, or with slight modification, as the opening paragraph of your report. Jewell's scope summary (Figure 11.2.33) reflects her initial decision to subordinate the failure investigation aspect and submit her report as a positive proposal to eliminate a series of personality problems.

11.2.4 Investigation/Preparation

At this point your development chart tells you why you are writing, to whom you are writing, and what your working title is. Your scope summary tells you what your product should be. You have classified lists of questions, the answers to which provide the material substance— what your audience needs and wants to know. You have only to dig to find the answers to the questions, then organize them, to be responsive to the assignment. All you need now is a search plan and facility to collect your acquired material.

Having arrived at your present position in the professional world, you need no further instruction on research, investigation, or other information-seeking methods. The advantage you now have is that before you leave your office, you have a firm idea of what you seek and how it will lead to your project solution. This is not to say that the investigation plan is fixed. Always be receptive to additional ideas or expanded areas in which to delve. Schedule your search routes efficiently to acquire the maximum amount of source material in the least time.

<u>Organizational Structure</u>

What is the organizational structure?
How does it compare with other service centers?
Those with Repair Supervisors?
Those without?
Why can't the present staff adjust to the need?
Will additional personnel be necessary?
Will the position be purely supervisory?
Does the present system need restructuring?
What are the alternatives?

<u>Cost</u>

How much would proposal cost?
How will cost be returned?
If direct return, in what ways?
If indirect return, in what ways?
Is there additional cost beyond the outlay of
    salary and overhead?
Adequate space?
Cost of desk and other facilities?
Where is person to come from?
New Hire?
Promotion from ranks?
Lateral transfer?
Should this person be permanent or temporary?
Will additional person "deflate" the job of
    testing foreman?

<u>Analysis Time</u>

What is the present RBA index?
What is the missed appointment rate?
What is the repeated report rate?
How do the indices and trouble indicators compare
    with other service centers?
What effect on computer usage and analysis time?
What is the coin trouble rate?

<u>Customer Service</u>

How will Repair Supervisor improve customer relations?
What is the percent (%) of customer takeovers?
How will Repair Supervisor reduce customer takeovers?

<u>Advantages</u>

How can additional person benefit the district?
How can additional person benefit the company?
How will additional person benefit the personnel?
Will personnel development and training increase?
Is there need for more development and training?
What effect on productivity?
How will overtime be affected?
How can person be used in the future?
Will this person continue to fit into corporate game plan?

(Total 41)

Figure 11.2.32 Primary sort: question categories for prototype project. The sorting is done on separate sheets. (Courtesy L. C. Jewell.)

```
┌─────────────────────────────────────────┐
│                                         │
│          SCOPE SUMMARY                  │
│                                         │
│                                         │
│    This proposal will solicit the addition of a │
│    management person to Moonrise Service │
│    Center's staff.  It will identify insufficiencies │
│    in the present staffing arrangement, and │
│    will describe   overall benefits to be │
│    gained from a force addition. │
│                                         │
│                                         │
└─────────────────────────────────────────┘
```

Figure 11.2.33 Scope summary: prototype project. (Courtesy L. C. Jewell.)

As you gather answers to your questions (data and readout sheets, photographs, sketches, written interview notes, photocopied book pages, lists, tear sheets, tables, etc.) you could very well collect them in a file box, and then waste a day or two sorting the lot into your groups. This sorting would then be followed by an inventory of the items. You can save yourself this monumental clerical task by using a simple device; there is a better way.

Use one 9 × 12-inch blank mailing envelope for each group. Mark each envelope label with the sublist title. You should have the same number of envelopes as you have question groups. As you gather answer materials, place each item into its appropriate envelope and note it on the label, in one, two, or three cryptic key words that best describe the item. In this way, you sort out and inventory the materials at the same time. This saves you time and provides a continuous inventory status.

Now you are fully armed; the thinking and planning stage is over. You can get out and attack your objective with confidence.

11.3   CONDUCTING THE INVESTIGATION

Conduct your investigation in whatever way your experience dictates: statically, by reading and using the observations of others; dynamically, by making your own observations. Accumulate as many answers to as many questions as you can in a reasonable amount of time. You do not have to obtain answers to the questions in the order they appear, and you do not have to number or index questions at this stage. Keep your scope summary at hand; try to get your answers so the material leads to your goal.

Keep two points in mind: try to be ethical in your quest; and avoid "one-stop shopping," seeking all your answers in one place. Store and inventory items as you acquire them, using the method described in Section 11.2.4.

You will find that you will not obtain answers to all your questions. Some may be impossible to answer; some may become moot as the search goes on. On the other hand, new questions may arise. Caution: Do not overextend the investigation. Stop when experience tells you "Enough!" You could choke to death on an overabundance of source material that may become unmanageable.

11.4 DATA DISTILLATION AND DISTRIBUTION

When you complete the field investigation, you can cross off most of the listed questions as answered. File the question sheets for possible later reference.

Photocopy all envelope faces. Use these copies as inventory face sheets for the rest of the process. The contents of the envelopes are your hard news, your "money in the bank." Store these in a safe place until you start writing. This material is listed on the inventory face sheets. Lay these sheets out on a table top, in console fashion. What you are looking at, in effect, is your report in its primordial form. When you organize these cryptic inventory key notations, you will have a real, firm outline. You need only add the detailed substance from the envelopes to complete your report. Figure 11.4 shows the five inventory face sheets that Jewell photocopied from the labels of her collection envelopes.

Major heading order is established by shifting the inventory sheets around to find the arrangement that most closely satisfies three

Figure 11.4 Inventory face sheets, photocopied from collection envelopes. (Courtesy L. C. Jewell.)

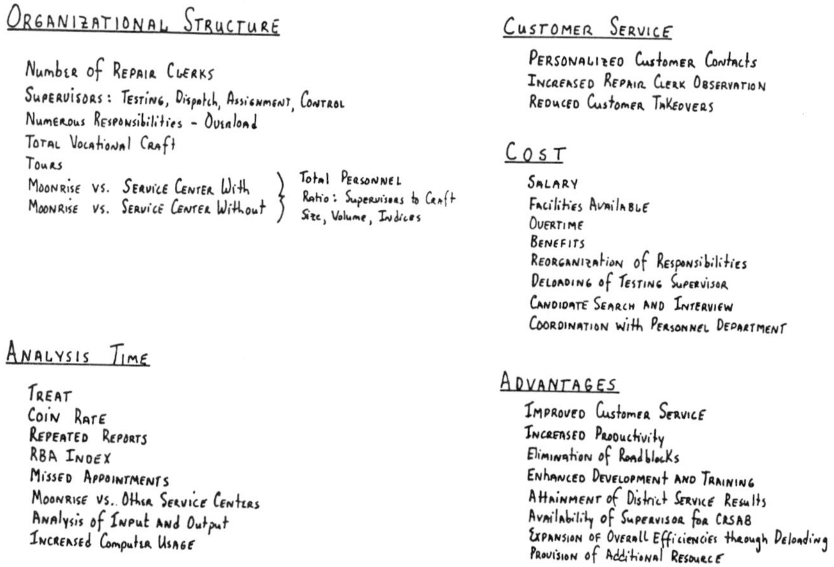

ORGANIZATIONAL STRUCTURE

Number of Repair Clerks
Supervisors: Testing, Dispatch, Assignment, Control
Numerous Responsibilities - Overload
Total Vocational Craft
Tours
Moonrise vs. Service Center With   } Total Personnel
Moonrise vs. Service Center Without   } Ratio: Supervisors to Craft
                                        Size, Volume, Indices

ANALYSIS TIME

Treat
Coin Rate
Repeated Reports
RBA Index
Missed Appointments
Moonrise vs. Other Service Centers
Analysis of Input and Output
Increased Computer Usage

CUSTOMER SERVICE

Personalized Customer Contacts
Increased Repair Clerk Observation
Reduced Customer Takeovers

COST

Salary
Facilities Available
Overtime
Benefits
Reorganization of Responsibilities
Deloading of Testing Supervisor
Candidate Search and Interview
Coordination with Personnel Department

ADVANTAGES

Improved Customer Service
Increased Productivity
Elimination of Roadblocks
Enhanced Development and Training
Attainment of District Service Results
Availability of Supervisor for CRSAB
Expansion of Overall Efficiencies through Deloading
Provision of Additional Resource

criteria: it is responsive to the charge in the original assignment; it reaches the expected goal as stipulated in the scope summary; and the writer is comfortable with it. Were Jewell to have done this by trial-and-error, she would have had to look at 120 possible combinations of the five sheets. Again, there is an easier way.

The first try is the way they fall when you lay them out, this is pure chance. Only once in all the years I have used this system did this first try become the final order.

When Jewell looked at her five groups, in the light of the three criteria, she arranged the sheets this way:

1.0 Advantages
2.0 Organizational Structure
3.0 Costs
4.0 Analysis Time
5.0 Customer Service

Then she realized that the root of the problem was a "people flaw" and tried again:

1.0 Organizational Structure
2.0 Customer Service
3.0 Analysis Time
4.0 Advantages
5.0 Costs

This seemed fine, but it left nowhere to go, and Costs was not sufficiently germane to the solution to stay a major group.

By culling several forward-looking notions from other categories, Jewell created a new group that would close on a strongly positive note, Future Needs. Then with some astute but honest reshuffling, she disbursed all the Costs items among the remaining groups wherever they best fit. Thus, Costs was eliminated as an integral heading, but the material was retained. To fill the void, Future Needs was renumbered as topic 5.0.

This reordering, a normal function of the overall operation, fit the three criteria reasonably well, although naturally several other arrangements were possible. Jewell fixed this order as the working datum, for the time being.

These rearrangements can be accomplished on a computer; however, it is better to do this primary sort with the actual inventory face sheets because you can shuffle them about on the table top and physically look at each combination to see if it makes sense.

Subsequent subordinated sorting will bring systematic order from the random positioning of the listed entries. Follow the next few steps in the procedure with great care because its product is your firm outline. Your document and, consequently, your project will stand or fail on the strength of this structure. Ostensibly when your outline is fixed, your document is finished. The reasoning is in place; the writing is just the facade.

11.5   OUTLINE BUILDING

To this point you have regarded the complete project as a monolith. Now the groups have come of age, so to speak. Thus you must deal with each group as a separate, but not isolated, entity. Groups must relate to each other as integral parts of a whole system. What you do individually with any group, you must do with all the rest in turn. This is a rapid development. It metamorphoses like a Polaroid Land photo. When the camera spits out the plate, you see a black surface (your inventory face sheet). As you watch, the blackness fades and quickly, before your eyes, a full color picture evolves (your firm outline).

The items listed under each major heading represent the information you have available. Some are important; others may go unnoticed in a more concise report. You arrange them in order of subordination and interdependence. The key is: How do these items relate to each other and to the whole? How do all points lead your real readers to the objective you set in your scope summary?

Try a simplistic example: The title is "The United States of America." The major group headings are 1.0   New England, 2.0   Middle Atlantic States, and so on. Secondary sorting leads to (states) 1.1   Maine, 1.2   New Hampshire, 1.3   Vermont, and so on. Tertiary sorting leads to (cities) 1.2.1   Concord, 1.2.2   Hanover, 1.2.3   Portsmouth, and so on. Later you might cross-index 1.2.3 Portsmouth with 2.3.2 Philadelphia if you were discussing naval bases; or 1.2.2 Hanover with 2.1.5 Ithaca, if the point were universities. Occasionally there are no intermediate entities to head subordinated items. Do not raise the lesser class listings to a higher level. It upsets parallel entries in another section. Instead, you may interpose inert umbrella headings as labels to classify subordinated entries.

In outline building always use the current, decimal-numeric indexing system, not the archaic, so-called Harvard system of unrelated Arabic and Roman numerals, capital and lowercase letters. The decimal system not only shows clearly the orders of subordination and interrelation, but also allows for cross-indexing and is useful in all computers.

As Jewell moved from her inventory face sheets (Figure 11.4) through the logical changes mentioned in Section 11.4, she followed the steps described earlier in this section, and composed the outline in Figure 11.51. For all general purposes, this order is satisfactory and would yield a respectable report. Though it met two criteria, she was not comfortable with it and thus could not write her best. Wanting a winner, she revised it to the firm outline in Figure 11.52 for the following reasons.

Although from the primary sort they emerged as separate major groups, both Customer Service and Analysis Time, for the refined overall project goal, would have to be made subordinate to an overriding tele-

1.0 ORGANIZATIONAL STRUCTURE

    1.1 Management to Vocational Structure
        1.1.1 Management
        1.1.2 Vocational
        1.1.3 Ratio
    1.2 Total Number of Stations
        1.2.1 Residence
        1.2.2 Business
        1.2.3 PBX

2.0 CUSTOMER SERVICE

    2.1 Customer Takeover Availability
    2.2 Improved RBA/Records Quality

3.0 ANALYSIS TIME

    3.1 Repeated Report Reduction
    3.2 Test Table Deloading
    3.3 Coin Analysis

4.0 ADVANTAGES

    4.1 Management to Subordinate Ratio
        4.1.1 Advantage to Testing Foreman
        4.1.2 Advantage to Repair Supervisor
        4.1.3 Advantage to Subordinates
    4.2 Training

5.0 FUTURE NEEDS

    5.1 ESS Cutover
    5.2 Centralized Repair Bureau

Figure 11.51 Preliminary outline option for prototype project report. (Courtesy L. C. Jewell.)

phone company consideration, Return of Costs. This notion always existed, but only in its component parts. They needed to be assembled for the entity to materialize. Thus, Customer Service and Analysis Time became parts of a superior idea. This is another reason for paying close attention to *why* and *to whom* you write, and how those considerations govern the entire process.

Future Needs stood up as a strong closing; thus this section remained intact. As she did earlier with Costs, Jewell now distributed the materials under Advantages among the other groups where each could prove its own point more potently than as part of a collection. Material was not discarded; it was only relocated to more effective positions.

Organizational Structure remained the opener because it clearly delineated the cause of the problem. The remaining minor difficulties appear jointly with Jewell's citation of each suggested remedial measure. This section mentions Analysis Time, but does not elaborate. To avoid redundancy, or wordage where it detracts from the focus, she cross-indexed this topic to another section, where it is detailed in a place more useful to the reader.

Lynn Jewell now decided that this firm outline (Figure 11.52) fit all three criteria discussed in Section 11.4. From it, and the stored reference

1.0 ORGANIZATIONAL STRUCTURE

    1.1 Management to Vocational Structure
        1.1.1 Management
        1.1.2 Vocational
        1.1.3 Testing Foreman to Subordinate Ratio
    1.2 Inefficiencies
        1.2.1 Personnel Development
        1.2.2 Meeting Objectives
        1.2.3 Flexibility
        1.2.4 Analysis Time (2.2.3)

2.0 RETURN OF COST

    2.1 Direct
    2.2 Indirect
        2.2.1 Personnel Development
            2.2.11 Advantage to Testing Foreman
            2.2.12 Advantage to Repair Supervisor
            2.2.13 Advantage to Subordinates
            2.2.14 Training
        2.2.2 Customer Service
            2.2.21 Customer Takeover Availability
            2.2.22 Improved Index
            2.2.23 Customer Report Rate
        2.2.2 Analysis Time (1.2.4)
            2.2.31 Repeated Report Reduction
            2.2.32 Test Table Deloading
            2.2.33 Coin Analysis

3.0 FUTURE NEEDS

    3.1 ESS Cutover
    3.2 Centralized Repair Service Answering Bureau

Figure 11.52 Firm outline for prototype project report. (Courtesy L. C. Jewell.)

material, she wrote a 32-page document that succeeded in its mission. It defined the trouble, identified the source, and suggested a satisfactory solution. Time has proved that her solution was correct because the problem ceased and has not reappeared.

When you have completed your firm outline, you are almost ready to write; however, don't rush into it. Many traditional teachers of technical writing advocate "cut-and-paste." In this process you write a draft of the report, then if some parts of the report appear unsuitable where they are, you physically cut those passages out, close and tape the gaps, then insert the misfit material where appropriate on other pages. If it still doesn't work in the new place, remove the peripatetic passage to yet a more potent position. The advent of the word processor, with its soft-lighted screen, cursor, and mouse, has made this practice so easy it is greatly overused, along with its enlarged flaws. It is not good. Why?

Whether by hand or by machine, each time you move a passage you create a lot of extra work to handle the aftermath. Any cross-indices would require rewiring. Then you must rework two transitions at the removal site, and again in four lines at the transplant, all to ensure flawless flow. If you take the time and care with your outline while it is in console display on your table top, you can see how all the parts fit the whole. *This* is the place to rearrange key thoughts, with their index numbers, and avoid the subsequent *shear folly.*

Unclear writing is a symptom of unclear thinking. If you organize your thoughts, your logic and reasoning will follow. Thereafter your source material provides your substance.

11.6   WRITER'S BLOCK

Writer's block may stop you from putting pen to paper at this point. Too many bright, well-educated people suffer pen paralysis, and surrender to the false notion they cannot write. This is an attitude, not a talent problem. Take the position that you have already won! Actually you have, with the completion of your preparation, investigation, and outline. The structural skeleton is in place; the filler is collected and inventoried.

To start properly you need an optimum environment— comfortable surroundings that provide an ambient atmosphere for easy writing:

An office to which you can close the door, or at least a workplace in which you can have emotional privacy

No outside noise (use earmuffs if necessary)

A comfortable chair to support good posture

Adequate lighting

A hard writing surface

Plenty of writing implements

All reference materials close at hand

Something to drink (do not nibble)

Eschew interruptions, discourage distractions, and give this project your total consideration.

Nowhere is it written that you must commence, "1.0 In the beginning." Your best start is with the section you know or like best. This enables you to handle the material with ease. In polishing off the draft easily, you both make progress and increase confidence.

The best strategy is to bypass the first and last sections and deal with the interior groups. Choose your "easy mark" to start, then write the rest of the sections, one at a time, in order of increasing difficulty. The decimal-numeric indexing system enables you to file your draft pages consecutively, even though you write them at your convenience. Write whole sections only, not parts or subsections. That only leads to fragmentation, then chaos.

After all interior segments are in first draft, write the closing. Then, with the full document at hand, you may write the opening, introducing an extant work. Because you have already written the report, you will not, in the introduction, mention items that are not discussed in the report body.

To summarize, suppose you have a six-section outline. Bypass sections 1.0 and 6.0. For this case only (because every document has a life of its own) suppose that of sections 2.0, 3.0, 4.0, and 5.0, you know section

5.0 best. Do that section first. Then write section 3.0, because it is the shortest and it flows into section 4.0. The remaining section, 2.0, is the hardest, but because of the experience gained by the time you reach it, it imposes no stress. Now section 6.0 will close logically; and with a drafted text in hand, the opening section, 1.0, may even be enjoyable. Your first draft is done!

Having selected the order of preference, you must now do the actual writing. Thanks to today's technology, there are many ways to get words onto paper. You may be skilled in the use of the word processor or typewriter; however, far fewer brain cells are required to push a pen over paper than to manipulate ten fingers on a keyboard. Because you need nearly total concentration on your expression, the proven best way to do your first draft is handwriting. Write on every other line of blank lined sheets, keeping before you the outline, the scope summary, and the envelopes containing source material.

Do not stop to look up a definition, spell a word, ponder punctuation, or mull over a phrase. Write as the notions come to you. Keep the thoughts flowing; everything goes into the first draft. Make no corrections or changes. Get it all on paper before the bird flies away. As you finish a section, file it by index number. This is the last of the piecemeal writing. From here on you handle the full document, in numerical order.

After the first draft is written, you can put away the envelopes and bring out the writing references: dictionary, thesaurus, grammar and spelling guides, style guide, and Bartlett's. You are ready to refine and polish.

You have three choices.

1. You could enter your first draft into a word processor. If your machine has been so programmed, it will correct your spelling and/or punctuation. You can rearrange, change words or paragraphs, edit, refine, and print out.
2. You can postpone word processing one step, and have another go at handwork. Review your full first draft using a different-color pen. Add missing words or lines, correct spelling and solecisms, then enter it into the word processor, or type it.
3. Some people speak better than they write. They are oral composers. For those people, the route is easy. Ready with reference material, outline, and scope summary, speak to the tape recorder as if you were speaking to the real reader. Transcribe the tape to hard copy, then edit. From there, the procedure is the same as for written efforts.

In Chapter 10, Hyzer discusses in depth the use of photographs as illustrations in forensic engineering reports. These, along with other selected graphics, should be incorporated into your report in addition to (not instead of) your narrative. Graphics augment text; they do not replace it. Each graphic should bear the number of the text section to which it refers. If possible, avoid numbering illustrations Figure 1, Figure 2, and so on, because this system does not really tell the reader anything, and removal of one figure requires renumbering of the remaining figures. In addition to the index number, include three notations in the caption: what the item is, what specifically to look for in it, and how it relates to the text. Do not forget to credit sources if needed.

One or two further passes may be required before the report is sent for reproduction. These efforts require patience and professionalism. There is no magic medicine of one dose.

11.7 A FEW HELPFUL GROUND RULES

These rules are not hard and fast, but are general guidelines that may help you write a better report.

Write in the active voice, first to second person. The archaic notion that technical writing must be in passive voice, in the impersonal third person, is no longer general practice. "We tested for you" is better than "tests have been made by this agency in behalf of the client."

Observe all the rules of English grammar, sentence structure, punctuation, and spelling, except when rigid adherence to a rule muddles the message. A bent rule will spring back; an obscured message could do harm.

Write short sentences, averaging about a dozen words. Should you need 47 words to explain an item clearly, use 47 words. Be sure it is not a run-on sentence. Occasionally, the opportunity arises to include a short sentence to break the monotony. Do.

Avoid complex or convoluted sentences. Be simple, but not simplistic.

Eliminate 15% of the fat from your effort by omitting dead-head (nonworking or redundant) expressions:
Cut "in other words" along with the preceding passage you feel compelled to rework.
Omit "in order" (or "so as") before any infinitive.
Do not use "located" before any form of the verb "to be."
Do not use "the following" (or "as follows") before a colon preceding a listing.
Use either "period" or "time," but not both.
If you cancel something, it *is* "out"; if you join two entities, they *are* "together"; if it is red, naturally it is "in color."
You know many other examples of unnecessary wordage. Think

of the savings of time and materials. If you omit them all, you will not lose one iota of your meaning.

Be positive, don't hedge. Your client pays you for definitive answers, not aphorisms.

Your name goes on this report. Your finished product must reward your efforts.

11.8 DELIVERY

The format you choose is purely cosmetic. The text is the substance that must reflect your forensic engineering work by being responsive to the charge in your original assignment. What you deliver to your retainer must meet the four minimum standards of receiver acceptability. Your report must penetrate the barrier of receiver resistance; start from the readers' correct knowledge datum and comprehension level; be sufficiently brief; and get the real reader involved.

11.9 WRITING

The culmination of a long chain of events, writing is a creative process wherein you cull facts from varied sources and throw them into your mental hopper. Then, in your own mold, you shape the resultant mash to be what you feel best expresses what your readers need and want to know. To some this seems impossible; to some it is a great struggle; to most it is a skill; and to a miniscule few it is a talent.

Remember you are working professionals, licensed products of institutions of higher learning. You have the mental capacity to make reasonable judgments on the basis of your observations. For you, writing is not impossible; nor should it be any more of a struggle than solving quadratic equations or balancing formulas. You just need the positive psychological attitude ("of course I can write"), the self-discipline that got you to where you are now, and these valid guidelines.

If you don't have to bother with any of this because you already command such a skill, then it will need honing every few years, because just telling it is not enough; it has to produce a finished product. And, the trouble with those rare, truly talented writers is their need for organization, a facet they do not believe they lack.

12. The Engineer as Expert Witness

JOSEPH S. WARD, P.E.

12.1 INTRODUCTION

In the preceding chapters of this book there has been considerable discussion concerning the investigative procedures undertaken by the forensic engineer, the development of professional opinions, and the preparation (if required) of the forensic engineer's written report.

All of the foregoing were accomplished with the assumption that the forensic engineer may eventually testify as an expert witness in legal proceedings. The details of the investigation, the gathering of data, the review of pertinent documents, the evaluation of data, and the conclusions and opinions developed by the forensic engineer are all undertaken with the view that the forensic engineer would eventually be sworn to give testimony in a court of law, in arbitration proceedings, or in other legal forums where direct testimony will be presented and the expert witness will be subject to cross-examination.

The posture taken by the forensic engineer in developing his or her hypotheses must always consider that the expert will be drawn into the legal process. But, in reality, this is not always the case. While the presumption exists that the dispute may eventually go to trial or arbitration, an overwhelming majority—over 95%–of the claims involving allegations of design professional error are settled out of court (Nelson 1987). Eugene I. Pavalon, Esq., President of the Association of Trial Lawyers of America and a member of the plaintiff's bar practicing in Chicago, has written:

> While lawyers have long estimated how few cases actually reach trial, the extensive "1983 Civil Litigation Research Project" of the University of Wisconsin Law School (CLRP) confirmed from empirical data that 90% of all citizen disputes were resolved by settlement without litigation. [Pavalon 1987]

In this writer's practice as a forensic engineer over the past 38 years, he has found that less than 10% of the cases in which he was involved ever reached the trial or arbitration stage. In more recent experience over the past 4 years, this writer has been retained as a forensic engineer in 62 cases and in this total caseload he has given testimony as an expert witness on two occasions in courts of law and in two arbitration hearings. This is about 6% of his caseload but it is acknowledged that some of the ongoing cases may eventually require expert witness testimony; however, the greatest percentage of disputes going to trial or arbitration are settled between the parties prior to ever reaching that stage.

Nevertheless in the life of every forensic engineer there will come a time when that individual will take his place on the witness stand and present expert testimony on the findings of the investigation. It is the purpose of this chapter to describe the posture of the forensic engineer when that individual becomes an expert witness.

This chapter is concerned with the resolution of conflicts in the construction industry by the conventional litigation procedures in courts of law and the role that the forensic engineer plays in the resolution of such conflicts. While this writer will explain the functions of the expert witness in court, the same philosophy applies to the expert when that individual testifies in arbitration hearings, although the latter may be somewhat less structured than normally required within the court system in the United States.

As a preamble to the expert appearing in court, it will be assumed that the forensic engineer will have completed the investigation and if an engineering report was rendered, that report will have been distributed to the opposing attorneys. It is also understood that all elements contained in the report will have been thoroughly discussed with the forensic engineer's attorney client. If a report was not prepared, the essence of the forensic engineer's findings will have been explained in detail to the client.

As an extension of the foregoing, it is inadvisable for the forensic expert to submit a draft report to the client with subsequent input from the attorney used to modify the final report submission. On cross-examination, the opposing attorney could make the inference that the report was not solely the work product of the forensic engineer and that his or her findings may have been influenced by the attorney client.

12.2   INTERROGATORIES

As the opposing parties prepare for trial, questions are asked of each other by the attorneys with responses to be furnished as mandated by court order. Interrogatories are a part of the discovery proceedings prior to trial.

It is the purpose of interrogatories to try to ascertain the legal positions of all parties to the controversy and to extract as much information as possible as to events, sources of data, names of individuals who have knowledge of the events that led to the dispute, to identify potential expert witnesses and to try to obtain the opinions of such experts, although in many instances such responses may be considered as attorney work product and therefore will not be answered by the respondents. However, if the experts have prepared reports and these reports have been distributed to the opposing parties, this objection may be irrelevant.

The forensic professional can be of assistance to the attorney client in the preparation of questions to be asked and in the phrasing of such questions to elicit information that will be helpful to the expert in either formulating his or her own opinions or for future use at trial.

In a similar manner, the forensic engineer can assist the attorney client in preparing answers to interrogatories based on the information that has been made available to the expert during the investigation.

In a sense, interrogatories are a "fishing expedition" by both sides to learn as much as possible about the positions of the parties prior to trial. At times, the responses may be evasive and may not be the information that the opposing attorney is seeking. This writer has found from his personal experience that usually only some of the answers to interrogatories are meaningful, but those that are can be extremely informative.

Discovery proceedings, such as interrogatories, are common in the usual pretrial procedure for litigation in courts. In arbitration proceedings, discovery is normally not permitted unless the parties to the litigation mutually agree to exchange information through interrogatories, depositions, or production of documents.

12.3 DEPOSITIONS

One form of discovery is for attorneys to take the deposition testimony of witnesses on the opposing side of the controversy. The deponents need not necessarily be professional experts, but can be fact witnesses to the event that precipitated the litigation. In the context of this chapter, discussion will concern only those professionals who testify as experts in the deposition proceedings.

The opposing attorney will issue a subpoena *duces tecum* through the court to be served on the proposed deponent. The subpoena will direct the witness to appear at a specific time, date, and location and will demand that the witness bring all of the documents contained in the expert's files to the deposition. The venue for the deposition is normally in the office of the opposing attorney. Present at such depositions, in addition to the opposing attorney, will be the attorneys representing other parties to the dispute (if they exist) as well as the attorney client of the witness.

Representatives of the principals in the litigation, as well as experts on the other side, may also be present. A stenographer will record all of the proceedings. In recent years, deposition hearings are frequently videotaped when this is agreeable to all parties.

Once an expert has been notified that a deposition will be taken, that individual should immediately contact the attorney client and thoroughly review with the attorney client all aspects of the case, including the anticipated questions and the manner and extent to which the witness will respond.

The initial questions to the witness will usually establish the witness' name and address, the name and location of the witness' office, the nature of the witness' practice, education, and professional background, and examples of projects in which the witness has been engaged, with specific emphasis on projects that relate to the matter of the litigation.

It is to be emphasized at this point that not only in deposition testimony but also in testimony in a court of law or in arbitration proceedings, all answers by the witness must be factual. The witness must *always* tell the truth. Not only is the witness sworn to "tell the truth the whole truth and nothing but the truth," but the veracity of responses to questions must never impinge on the witness' credibility. However, in responses to questions in deposition, and in responses during cross-examination in court or in arbitration, the witness should limit answers only to the questons asked and should not elaborate any further in testimony.

The questioning by the opposing attorney will continue with specific aspects of the witness' involvement in the matter of the litigation. If necessary, the witness may consult with the attorney client during the course of the questioning. If an objection to a question is raised by the attorney client, the witness should not respond at that time but should do so only when the attorney client tells the witness to answer the question. The objections will be noted on the record and may later be the foundation for an objection if the transcript of the deposition is read at the trial. The attorney client may deem that answers to certain questions are attorney work product and therefore are not admissible in the deposition proceedings. The attorney client may instruct the witness to refrain from answering such questions.

Following the questioning by the opposing attorney who called for the deposition testimony, questions may be asked of the witness by other attorneys present. Cross-examination questions may be posed by the attorney client which may be followed by other questions by the attorneys in the room.

The demeanor of the witness should be professional at all times. The witness should act in a manner that will enhance credibility in the interrogation. Remember, all of the witness' testimony is being recorded and can be used at a future trial.

If the witness does not know the answer to a question, he or she so states. If the question is not clear, the witness asks for the question to be repeated or for a clarification of the question, which may be done by the opposing attorney, or the question may be read back by the reporter.

"Body language" by the witness in responding to questions can be important, as the impression of an unsure response may lead the opposing attorney to delve into more aspects of the witness' testimony. This is particularly critical when the deposition is videotaped and is presented at trial in the future.

Deposition hearings for a single witness can consume a considerable amount of time. It is not unusual for a deposition to last an entire day, or several days, and it has been known that the hearings can be protracted over a period of weeks or months. This depends not only on the detail of the questions asked by the opposing attorney but by the queries of other attorneys present and the cross-examination by the attorney client. In a multiparty litigation, frequently many attorneys, and their clients and experts, fill the room and each of the attorneys propounds numerous questions of the witness. This writer has recently experienced the taking of his deposition where the number of parties involved in the litigation required the presence of 15 attorneys (all who had something to ask) and the deposition consumed 5 days of testimony.

Deposition testimony can be one of the critical elements in pretrial discovery and cases may be either won or lost by deposition testimony, frequently resulting in an out-of-court settlement. Although at times the adversary attorney may appear to be pleasant and accommodating, he or she is involved in the litigation to win for his or her client. The witness should not become careless and be fooled by innocuous inquiries; they may be the prelude to a salient question that the opposing attorney may ask with the hope that the response will discredit the witness' testimony.

When depositions are recessed either during the day or at the close of the day, the witness should consult with the attorney client to review the witness' responses and possibly to rehabilitate testimony through cross-examination by the attorney client when the session resumes.

At the conclusion of the deposition, the witness will be asked if a reading and possible correction of the deposition transcript can be waived. This should be rejected as it is usually a good practice for the witness to review testimony when it is transcribed and to correct improper recordings by the stenographer. The essence of the witness' testimony, however, may not be changed.

In some jurisdictions it is permissible for an expert retained by an attorney client to audit the depositions of experts retained by the opposing attorneys. Where this is allowed, and is agreed upon by all attorneys in the litigation, the forensic engineer may sit with the attorney client and give advice as to questions that should be asked of the witnesses whose depositions are being taken. This is usually in the form of handwritten notes conveyed by the expert to the attorney client that help to elicit technical and other information from the opposing expert. Keep clearly in mind the importance of deposition testimony. It is a crucial part of our current litigation system.

12.4   TRIAL OR ARBITRATION PREPARATION

When it has been established that the case will go to trial or arbitration, much work is to be done by the expert and by the attorney client to prepare for the eventuality that the forensic engineer will take the witness chair. The writer was once told by an attorney in his early days of testifying as an expert witness that the secret for the attorney winning a case is by "preparation, preparation and more preparation." This is as true today as it was in the early 1950s.

Throughout the following discussion, the term *trial* refers to the court system but the same procedure can be utilized in arbitration proceedings.

The expert may have prepared an engineering report pertaining to his or her findings and opinions. If so, and if the report is to be used at the trial, the exhibits contained therein should be prepared to be introduced as evidence in the trial during direct examination of the witness. If the exhibits were originally of small size, and if pertinent photographs were used in the report, they should be enlarged such that they can be observable by the jury, the judge, and/or the arbitrator(s). If a report was not prepared in advance of trial, large exhibits may be made for the same purpose. If additional exhibits are needed to clarify technical testimony by the expert during testimony, they should be made in a similar manner.

Any exhibits produced at trial, whether they be the expert's report, reports of others, clarification drawings, photographs, or other items, will be initially offered by the attorney client for identification and, if they are acceptable to the court, they will be entered as evidence. If they are not entered in evidence they cannot be referred to by the witness in any testimony. Therefore, exhibits should be carefully selected such that the court can overrule any objections raised by the opposing attorney.

All aspects of the case in which the forensic engineer has been involved should be thoroughly discussed with the attorney client well before the trial date is set. In these discussions, there should be a rehearsal of the direct testimony to be given by the witness. It is often desirable to prepare an outline of the direct testimony sequence. The attorney client may even go so far as to write down the questions to be asked of the witness, so that the attorney client and the expert can evaluate the content of the witness' response.

In a similar manner, the forensic engineer should confer with the attorney client on the questions anticipated by the opposing attorney in cross-examination and both the expert and the attorney client should evaluate the expert's responses to those questions.

In introducing exhibits for identification, and eventually as evidence, the expert and the attorney client should consider the foundation that should be laid before the witness testifies as to utilization of the exhibits in testimony. For example, the witness may use photographs in the course of testimony. If such photographs were taken by others, whether or not in the presence of the expert, the photographer, a fact witness, should precede the testimony of the expert. These photographs will then be entered into evidence, with eventual testimony to be given by the expert on their merits. Similarly, if field or laboratory data were developed by other than the witness, those who obtained the data should be called to represent the integrity of the data prior to the witness' testimony. Therefore, a sequence of witnesses, whether they be factual or expert, should be developed in order that the exhibits required by the witness will be substantiated prior to the time the expert takes the stand. Frequently this requires a meeting of all witnesses to be called by the attorney client to prepare, in an orderly fashion, the introduction of testimony culminating with the laying of a proper foundation for the testimony of the expert witness.

It too often happens that prior to the testimony of an expert, a brief consultation is held with the attorney client in the corridors of the courthouse. If this is the case, a disaster can ensue when the expert takes the witness chair. Thorough preparation is essential.

The preparation for trial in a court of law is no different from the preparation needed when a witness testifies in an arbitration hearing. Although the procedures are more relaxed in arbitration and the rules of evidence may not be as stringent as in court, preparation for the witness' testimony should be just as comprehensive.

12.5 THE TRIAL

Frequently a settlement is reached between the parties before the trial commences. This may be the result of a pretrial conference with the judge, where the justice essentially acts as a mediator in trying to induce the parties to settle without going to trial. In other instances, during the pretrial discovery period, it may become evident to one or more of the parties that it would be disadvantageous to them to embark on a possibly long and costly confrontation in court, leading to proposal of a settlement. In arbitration proceedings, where discovery is not normally employed, the parties to the dispute may agree to settle prior to the commencement of the arbitration hearings.

As often happens, the trial or arbitration may begin and, when an attorney hears the testimony of the opposing expert, the attorney may call for a settlement conference. It has been previously recited that only about 10% of the cases eventually go to trial or to arbitration. More will be said in

Chapter 13 on alternate dispute resolution procedures that replace litigation in the courts. Paramount among these procedures is a negotiation or mediation process that has worked successfully in avoiding the courtroom.

A trial date is set in the appropriate court and all parties prepare for the trial. However, delays in trial commencement are frequent. The assigned judge may have underestimated the time that has been taken to conclude a prior case, or attorneys or witnesses are not available because of legitimate schedule conflicts, and the attorneys petition the court to delay the trial date.

Similarly (and possibly to a greater extent), the scheduled date for the commencement of arbitration hearings is postponed because one or more of the arbitrators, the attorneys, or the witnesses are unavailable. The writer has seen, in his many years of forensic engineering experience, scheduled trial or arbitration dates delayed by weeks or months.

Because of this, the witness who is to testify must remain available at the pleasure of the court, the arbitrators, and the attorneys on both sides. For example, while a trial date may be set by the court, the jury selection process, the time required for testimony by preceding witnesses, or recesses by the court, can hold the expert in readiness for a protracted time until he or she takes the witness chair. This can be aggravating and time consuming to the expert who must remain ready for the telephone call from the attorney client advising the expert to appear in court. And it does not stop there. The attorney client may have assumed the appropriate time for the witness to take the stand but preceding testimony may require the proposed witness to "warm his or her heels" in the courtroom, or in the courthouse corridors if the court or the jurisdiction does not allow the witness to be present during the testimony of others. This is a frustrating experience, but is all too common, and the forensic engineer should expect this possible eventuality.

12.6   DIRECT EXAMINATION

The expert has taken the witness chair. The expert is sworn by the clerk of the court and direct examination by the attorney client commences. Several points should be noted here prior to and during the expert's testimony.

The witness should dress conservatively and should display the aura of a professional. The witness should speak clearly and concisely and answer directly the questions asked by the attorney client, the opposing attorney, and, on occasion, the judge. Although the questions will be posed by the attorneys, if the trial is by jury, the witness should speak directly to the jury; if it is a nonjury trial, the witness' responses should be directed to the judge. Even in a jury trial, the witness should occasionally address answers to the judge.

Always avoid an argument with the opposing attorney. Although that individual may try to "rattle" the witness by adopting a demeaning manner toward the witness to provoke an argumentative situation, the witness must always "keep cool" and respond as a professional. Remember, the *witness* is the expert, not the opposing attorney. This writer has always taken the position that as a technical expert, he knows more about engineering technology than the opposing attorney whose technical knowledge may be limited to that gained from discussions with his or her own experts.

With the foregoing as a preface, the attorney client commences direct examination of the witness. The initial purpose in this examination is to qualify the witness to the court as an expert. Until such time that the witness is so qualified as an expert, the forensic engineer on the witness stand is simply a witness and cannot render an opinion. In the process of qualification, the attorney client will elicit from the witness information regarding the witness' education and professional background, history of employment, and prior projects on which the witness has worked. The witness will be asked to recite primarily those projects that have a bearing on the issues in the case under litigation.

After the witness properly responds to the attorney client's questions, the attorney will then move to have the witness qualified as an expert. The opposing attorney may request a *voir dire* of the witness, which is basically a cross-examination on the witness' credentials, with the hope that the opposing attorney can prevent the forensic engineer from being qualified as an expert. If the opposing attorney is successful in discrediting the qualifications of the witness to the extent that the witness cannot be considered an expert, the forensic engineer can only testify as to facts and cannot render an opinion. If the court rules that the witness is truly an expert, the direct examination continues in accordance with the questioning prepared in advance of trial.

At that point in the direct examination when the expert witness' testimony is to refer to exhibits, the attorney client will introduce these for identification. The opposing attorney may then *voir dire* for the purpose of not allowing the exhibits into evidence. If this endeavor is successful, the witness may no longer testify as to that exhibit. If the court overrules the opposing attorney's objection, the direct examination continues.

On occasion, the attorney client may ask the witness a "hypothetical question." In this procedure the attorney client will state a series of assumed facts and will ask the witness to give opinions on the basis of these facts. The hypothetical question sometimes is raised when there is insufficient direct factual evidence in the case to support an expert's opinion. The attorney client will later draw a corollary to the case in litigation to prove a point for the attorney's client. The hypothetical question presentation can be dangerous if it is not rehearsed well in advance of the trial. The expert should have full knowledge of the hypothetical questions to be asked and he or she should determine responses to these inquiries. The opposing attorney may object to the presentation of the hypothetical question and if the objection is sustained, the attorney client has lost the point. If the opposing attorney is overruled, the attorney client will utilize the testimony in the final summation to the court.

When the basis for an opinion by the expert witness has been firmly established through direct examination, the attorney client will ask the witness to give an opinion on the evidence and facts presented. A usual follow-up question by the attorney client will inquire of the witness as to how he or she arrived at that opinion.

At this point, the witness becomes a *teacher*. The witness will take all the time that is needed to explain, in simple language, the basic engineering fundamentals, together with the facts and evidence at hand, to substantiate the opinion that has just been rendered. When exhibits have been entered into evidence the witness may request permission to approach the exhibits on the floor and, maintaining eye contact with the jury and the judge, will explain how the exhibits influenced his or her opinions. The witness may also request permission of the court to develop, on flip charts, sketches or other presentations that will augment the testimony. Although such courtroom graphics may raise an objection by the opposing attorney, usually the court will rule that this is permissible for the expert to develop his or her rationale.

The testimony of the expert in developing a hypothesis could be quite lengthy and the witness should not try to hurry the explanation at the expense of not having the judge or the jury fully comprehend the basis for opinions. This writer has "lectured" to the court in the substantiation of his opinions a number of times, and, in a recent case, the court presentation consumed the better part of a day.

On conclusion of the witness' testimony, the attorney client will usually state that he or she has no more questions and the court will ask the opposing attorney to commence cross-examination.

A few points to remember during direct examination:

> The witness may use notes and data brought to the witness stand but the witness should refrain from reading from such data. The data should be used only to recall memory as to dates, numbers, and other pertinent information that will be used in testimony. Reading from such notes and data could present a problem if the opposing attorney raises an objection and requests to see these documents; or they may be requested by the opposing attorney during cross-examination. The documents could contain information that should not be in the hands of the opposing attorney. This writer generally uses a set of dates and numbers on a plain sheet of paper that he has brought with him to the witness stand. This brief information is the only material that the witness should have while providing testimony. However, at the time that exhibits are entered into evidence, the witness may freely have these exhibits during testimony.

Avoid testimony on hearsay evidence. Hearsay is prohibited under the court Rules of Evidence. As a witness, speak only from personal knowledge and not from what someone has said, unless the subject matter has previously been presented at the trial.

Stop talking if objections are raised by the opposing attorney, either to the witness' attorney client's question or to the response of the witness, until this objection is ruled on by the court.

12.7   CROSS-EXAMINATION

Although the expert witness has had all the time that was needed to make a detailed presentation during direct examination, the forensic engineer now faces the fact that most, if not all, of the testimony will be challenged by the opposing attorney. It is often said that cross-examination "separates the men and women from the boys and girls."

The opposing attorney will do everything possible to discredit the testimony of the expert witness. An attempt will be made to have the witness admit uncertainty of specific facts, that the analysis was incorrect, that assumptions were improper, that the witness did not have sufficient educational and practice background to properly analyze the controversy, and that opinions were invalid.

The opposing attorney will do this by asking pointed questions of the witness. The attorney may read from the witness' prior deposition testimony to develop conflicts in the expert's testimony. On occasions, the attorney will badger the witness, although this tactic will elicit an objection from the attorney client, which may be sustained by the court.

A good opposing attorney will tend to "put on a show" to impress the judge and the jury that the testimony presented by the expert witness is invalid. This procedure will be used to discredit the witness' testimony in the opposing attorney's final summation.

The expert witness should always refrain from being argumentative with the opposing attorney and should consistently display a professional attitude in responses to questions.

The posture of the expert witness should be similar to that which was displayed during a deposition, as previously explained in Section 12.3.

The following are some guidelines for the expert witness during cross-examination:

The witness should answer questions briefly without elaboration.

If the witness does not know the answer to a question or does not understand a question, the witness should so state.

If the attorney client raises an objection, the witness should not respond to a question until a ruling from the court is obtained.

The expert should not rush responses but should think clearly before answering the questions of the opposing attorney.

The expert witness, a professional, should not get "rattled." This writer always advises that witnesses "keep their cool."

The witness should not be concerned about misleading impressions that may have been created in responses to questions. These usually can be clarified on redirect examination by the attorney client.

Occasionally, the witness may be asked if he or she has been "paid for the testimony." The witness should always respond to such questions by stating that a fee has been paid for the services performed, and will be paid for the expert's presence in court. If asked the amount of the fee or the basis for the charges, the witness should always answer truthfully and directly.

With respect to the last point, the forensic engineer should never be engaged in a case where the fee is determined by the outcome of the litigation. This is commonly referred to as a contingency fee and, although it is frequently used in the legal profession, it is inappropriate for the engagement of a forensic engineer. The attorneys in the case are advocates for their clients. The forensic engineer is not. The expert has arrived at an opinion based on facts and should always be unbiased with respect to the issues of the case. Advocacy should always be avoided by the expert and this impression should be clearly portrayed to the judge and the jury.

12.8 REDIRECT AND RE-CROSS-EXAMINATION

At the conclusion of the cross-examination, the attorney client will resume questioning the witness, primarily to enable the expert to clarify responses to the questions raised by the opposing attorney during cross-examination. The areas pursued by the attorney client usually will be on the same lines as developed during direct examination and on subject matter raised during cross-examination. If new subject matter is brought up by the attorney client that was not covered in the direct or cross-examination, objections raised by the opposing attorney may be sustained by the court.

After redirect examination, in re-cross-examination the opposing attorney may wish to challenge the witness' responses in redirect examination, in a further effort to discredit the expert's testimony. The posture taken by the witness should be similar to that exhibited during cross-examination.

The redirect and re-cross-examinations may continue by the attorney client and the opposing attorney, respectively, until no further questions are asked of the witness and the witness is asked by the judge to step down from the witness chair.

12.9 EXPERTS PRESENT DURING TESTIMONY OF OPPOSING EXPERTS

Some courts will permit the expert who will testify or has testified to be present in the courtroom during the testimony offered by experts retained by the opposing attorney. If this is permitted before the expert testifies, this will enable the expert to ascertain the nature of the technical testimony introduced by the opposing side and also may permit the witness to discuss that testimony with the attorney client during a recess. The opportunity may exist to develop questions for the expert to contradict the testimony offered by the opposition.

Frequently, some jurisdictions do not permit the expert to be present in the courtroom during the testimony of other witnesses. In that event, during recess, the attorney client can discuss with the expert outside of the courtroom the testimony that has been offered by others. In addition, the attorney client can develop with the expert additional questions that will be asked of the witness on direct examination. If the opposing testimony is already in progress, the expert can advise the attorney client on pursuing a line of questioning in cross-examination.

Some jurisdictions even go beyond forbidding the presence of an expert in the courtroom. In a recent case in Delaware, this writer was admonished by the judge that in recess during direct examination, he was not permitted to discuss his prior or future testimony or any other aspects of the case with his attorney client.

12.10 REFERENCE DOCUMENTS

The Interprofessional Council on Environmental Design (ICED) has recently adopted the *Recommended Practices for Design Professionals Engaged as Experts in the Resolution of Construction Industry Disputes* originally proposed by ASFE (ICED 1988). The document is reproduced in the Appendix. To the knowledge of this writer, this document will be the only generally accepted published document to date that sets the criteria for the posture that should be taken by the expert involved in litigation matters. It has been endorsed by several of the major engineering and architectural societies.

Another document will be published in 1989 by the American Society of Civil Engineers through its Technical Council on Forensic Engineering. This document, *Guidelines for Failure Investigation*, covers detailed aspects of forensic investigations related to failures in the construction industry. It also dwells on the services rendered by forensic engineers when testifying as expert witnesses [ASCE/TCFE in preparation).

An excellent publication is one produced by ASFE, *Expert: A Guide to Forensic Engineering and Services as an Expert Witness* (ASFE 1985).

Four other publications may be of interest to the forensic engineer serving as an expert witness. One has been produced by the Defense Research Institute (DRI): *Operator's Manual for a Witness Chair*. This manual is concerned basically with a witness testifying at a deposition (Baker 1983). Another publication produced by DRI is *A Legal Primer for Expert Witnesses* (Postol 1987). *The Consulting Engineer as an Expert Witness* was developed by the Consulting Engineers Association of California (CEAC 1983). Finally, reference is made to an article published in *Better Roads*: "Ten Commandments (More or Less) for the Expert Witness" (Pagan 1982). In addition, this writer has presented a one-day course on forensic engineering for the ASCE Continuing Education Division on twelve occasions to date, at locations throughout the United States. Contact for future course presentations may be made through ASCE headquarters in New York City. Other references have been cited in the bibliography to this chapter to give the reader additional information on the role of an expert witness.

12.11 CONCLUSION

In conclusion, this writer offers some final thoughts in testifying as an expert witness:

Do a thorough investigation.

Be convinced of the facts obtained.

Keep the attorney client appraised of the forensic engineer's findings.

Do intensive preparation for testimony at deposition, trial, or arbitration.

As an expert on the witness stand, always act in a professional manner.

Be a "teacher" in presenting testimony before a judge, a jury, or the arbitrator(s).

As an expert witness, be "believable."

This chapter describes the posture of the forensic engineer when serving as an expert witness. Appearing in this capacity is not for the timid or the neophyte engineer. It requires a knowledge of the legal process and many years of engineering experience to testify in court or in arbitration proceedings in a manner such that the witness' testimony will evidence credibility and will be convincing to the court, jury, or panel of arbitrators. The testimony should stand up under scrutiny and should not be destroyed by the opposing attorney during depositions and under cross-examination.

Above all, the expert witness must be unbiased and objective in both the investigation and in testimony on the witness stand. The witness must never be considered an advocate, and should always "call the shots as they are." Expert witness testimony is an exciting endeavor and is one that is aimed at serving the cause of justice.

References

Aidlin, S. S. 1952a. The expert witness prepares for court. In *Professional Engineering—Economics and Practice*. Brooklyn, NY: Stephen Howard Co.

Aidlin, S. S. 1952b. The engineer as an expert witness. In *Professional Engineering—Economics and Practice*. Brooklyn, NY: Stephen Howard Co.

American Society of Civil Engineers, Technical Council on Forensic Engineering (ASCE/TCFE). *Guidelines for Failure Investigations*. New York: ASCE.

Association of Soil and Foundation Engineers (ASFE). 1985. *Expert: A Guide to Forensic Engineering and Service as an Expert Witness*. Silver Spring, MD: ASFE.

Baker, T. O. 1983. *Operator's Manual for a Witness Chair. Model 1: Deposition Chair*. Milwaukee, WI: Defense Research Institute.

Bleyl, R. L. 1985. The forensic engineer on the witness stand under cross-examination attack. *Journal of the National Academy of Forensic Engineers*, (NAFE, Alexandria, VA) 2, No. 2 (December).

Cantor, B. J. 1966. The expert witness. *American Bar Association Journal*, (Chicago) (October).

Cantor, B. J. 1987. *The Expert Witness*. Belmont, MA: Civil Evidence Photography Seminars.

Carper, K. L. ed. 1986. *Forensic Engineering: Learning from Failures*. New York: ASCE.

Consulting Engineers Association of California (CEAC) 1983. *The Consulting Engineer as an Expert Witness*. Burlingame, CA: CEAC.

N. Charlmers Inc. 1983. *On Being an Expert Witness*. Teaneck, NJ: Nelson Charlmers, Inc.

Compressed Air. 1986. The forensic engineer. *Compressed Air* (Ingersoll-Rand Co., Washington, NJ) 91, No. 12 (December).

Hanley, R. F. Working the witness puzzle. *Litigation* (American Bar Association, Chicago).

Huston, J. 1979. Engineers on the witness stand. *Civil Engineering* (ASCE) 49, No. 2 (February).

Interprofessional Council on Environmental Design (ICED). 1988. *Recommended Practices for Design Professionals Engaged as Experts in the Resolution of Construction Industry Disputes*. Silver Spring, MD: ICED (through ASFE).

Kornblum, G. O. 1974. The expert witness and consultant. *The Practical Lawyer* 20 (March):13–34.

Mazur, S., W. J. Postner, R. A. Rubin, and J. S. Ward. 1985. The engineer as an expert witness. Audiotape from ASCE Continuing Education Services, ASCE, New York.

Mazur, S., and R. A. Rubin. 1977. *Using Experts in Civil Cases*. New York: Practising Law Institute.

McElhaney, J. S. 1978. Trial notebook—an outline on hearsay. *Litigation* (American Bar Association, Chicago) 4, No. 2 (Winter).

McQuillan, J. A. 1984. The C.E. as an expert witness. *Consulting Engineer* (Technical Publishing, Barrington, IL) 62, No. 1 (January).

Nelson, S. L. 1987. ADR—A different ritual: An insurer's perspective. *Journal of Performance of Constructed Facilities* (ASCE) 1, No. 4 (November).

Pagan, A. R. 1982. "Ten commandments (more or less) for the expert witness." *Better Roads* (Park Ridge, IL) (March).

Pavalon, E. I. 1987. ADR—Trial lawyer's perspective. *Journal of Performance of Constructed Facilities* (ASCE) 1, No. 4 (November).

Postol, L. P. 1987. A legal primer for expert witnesses. *For the Defense* (Defense Research Institute, Chicago) 29 (February).

Pritzker, P. E. 1984. Investigative techniques. *Consulting Engineer* (Technical Publishing, Barrington, IL) 62, No. 1 (January).

Richter, I., and R. S. Mitchell. 1982. *Handbook of Construction Law and Claims*. Reston, VA: Reston Publishing Co.

Specter, M. M. 1984. Engineering applied to jurisprudence. *Consulting Engineer* (Technical Publishing, Barrington, IL) 62, No. 1 (January).

Specter, M. M. 1987. National Academy of Forensic Engineers. *Journal of Performance of Constructed Facilities* (ASCE) 1, No. 3 (August).

Ward, J. S. 1985. Forensic engineering. *ASCE Continuing Education Lecture Notes* (September).

Ward, J. S. 1986. What is a forensic engineer? In *Forensic Engineering: Learning from Failures.* New York: ASCE.

Ward, J. S. 1986. Failures: Who did it—How did it happen? Proceedings of the Symposium "Management Lessons from Engineering Failures," ASCE, New York. NY.

Ward, J. S. 1987. Role of the engineering firm. In *Hazardous Waste Disposal and Underground Construction Law,* Chap. 6. New York: John Wiley and Sons.

Worrall, D. G. 1984. "Engineer experts—The attorney's viewpoint. *Journal of the National Academy of Forensic Engineers* (NAFE, Alexandria, VA) 1, No. 2 (October).

13. Role of the Forensic Expert in Dispute Resolution

JOSEPH S. WARD, P.E.

13.1 INTRODUCTION

The role of the forensic expert in dispute resolution in a court of law or in arbitration proceedings has been described in Chapter 12. There are other forums where the forensic engineer can play a vital part in assisting in the resolution of conflicts in a far more expedient manner. Such forums that attempt to avoid the traditional approach to litigation are known as alternate dispute resolution (ADR) procedures.

The theory behind the use of ADR is that a conflict is most easily solved immediately after the event that precipitated the conflict has arisen, rather than through the long and costly route of traditional litigation or arbitration.

Historically, claims emanating from a conflict may be filed well after the event has occurred. The claim initiates the traditional process that involves the legal profession from the outset, where attorneys are retained as adversaries on behalf of the plaintiff and the one or more defendants.

In the typical case, the following scenario then develops. The attorneys for the defendants file an answer to the complaint and may also file a counterclaim. The plaintiff's attorney will answer the counterclaim and may also produce an amended complaint. Third parties may be brought into the litigation by the plaintiff or the defendants and claims will be made against these third parties, with the ensuing answers. There will be a demand by all parties for the production of documents, and interrogatories will be filed, with the expected answers to such interrogatories. Expert and fact witnesses will be retained by attorneys for the various litigants who will conduct their investigations. The experts will produce reports that will be distributed to all parties involved in the conflict. As part of the discovery proceedings, witnesses may be subpoenaed for deposition testimony. Motions will be made to the court by the attorneys, such as to dismiss the complaints, and the courts will rule on these motions.

Finally, a court date will be set. However, with the current crowded court calendars in most jurisdictions throughout this country, a year or more can elapse before the court date is fixed, and this may be further extended by the court as described in Chapter 12.

At any time during the foregoing process the parties may agree to settle their differences, and a settlement agreement may be reached. But even when the parties decide to settle, a long time usually has elapsed since the occurrence of the event that initiated the conflict. If a settlement is not reached before the trial date, the court proceedings commence, and if settlement is not attained prior to the end of trial, a protracted period will ensue because of the testimony of witnesses and the pleadings of the attorneys.

In this writer's experience, it is not unusual for a dispute litigated in the courts to take years before it is resolved. In a paper in the November 1987 issue of the ASCE *Journal of Performance of Constructed Facilities,* Sandra L. Nelson (Vice President of Engineers Risk Services, a company that manages professional liability insurance programs) states that "The average life of a design professional liability claim file is still between three and four years, an improvement over statistics from a year earlier" (Nelson 1987).

The cost of the traditional litigation procedure is astronomical and has been accelerating in recent years. These costs are accrued by all parties to the dispute, primarily through the incurred legal fees, as well as for the time expended by the employees of the plaintiff and the defendants. The final award made to the aggrieved party will be a long time in coming. Even when damages are assessed and an award is made, it has been reported that an average of 70% of that award goes to the legal costs incurred during this entire procedure.

Although the foregoing describes current litigation in the courts, the same may be said, perhaps to somewhat or a lesser degree, when the dispute is resolved by arbitration. Even if discovery is eliminated in arbitration proceedings, the time elapsed between the demand for arbitration and the resolution of the dispute may consume months, or even years. The cost incurred by all parties to the dispute can be extremely high.

As an example, the American Arbitration Association (AAA) has reported that during the period January 1, 1986 to December 31, 1986 for the construction industry, "the overall average time from filing to award was 238 days" (AAA 1987). With reference to that same period, claims under $10,000 were awarded within 137 days on the average and those between $500,000 and $1,000,000 took 515 days on the average to conclude.

In his paper in the November 1987 issue of the ASCE *Journal of Performance of Constructed Facilities*, Eugene I. Pavalon (President of the Association of Trial Lawyers of America and a member of the plaintiff's bar practicing in Chicago) states that a study (Lyons 1985) has shown that 400 AAA construction disputes took an average of 6.5 months from filing to completion (Pavalon 1987). As to costs, Pavalon recites (Lyons 1985): "I tell my clients that it should cost just as much for a complex construction arbitration as it costs for litigation." Thus it can be seen that elapsed time and associated costs are high for dispute resolution in the courts and in arbitration.

There must be a better way to resolve such conflicts in a timely manner and at the least possible cost to the participants in the dispute. One answer, in addition to court reform, is to utilize several available methods of alternate dispute resolution.

The forensic engineer can be successfully involved in any of the several forms of ADR. The forensic expert, as a consultant to one of the parties involved in a dispute, can recommend suggested ADR procedures for expeditiously resolving potential conflicts and can advise clients not to pursue the traditional litigation process in the courts until all prior avenues for resolution have been exhausted. In certain forms of ADR, the forensic engineer can be a participant to enable the parties to the dispute to settle their grievances in the least possible time at minimum cost.

13.2   NEGOTIATION OF DISPUTES

In his paper which appeared in the same November 1987 issue of the ASCE *Journal of Performance of Constructed Facilities*, James W. Poirot (Chairman of the Board of CH2M Hill) stated the following:

> With people of high integrity and a renewed commitment to quality and proper attitudes, disputes can be resolved without resorting to formal dispute resolution techniques. If formal ADR techniques are necessary, the parties must commit themselves to resolving disputes amicably if possible—the only alternative is expensive litigation. [Poirot 1987]

Poirot further states:

> Once a project is under construction, all parties should recognize that potential disputes lie around every corner. Changed conditions during construction should be recognized as normal, not as unusual circumstances.

> If a dispute does occur, all parties should develop an attitude of compromise with a recognition that a quick resolution of the dispute and a continuing harmonious team will be to the benefit of the project.

> If the team members cannot resolve the changed conditions and a true dispute develops, the first attempt at resolution should be by

internal negotiations. This requires a meeting of those in authority to quickly resolve the dispute.

Successful negotiations are most often win–win solutions. That is, both parties feel they have compromised but, nevertheless, have been a winner in the negotiating process.

The preceding discussion refers to the parties getting together, without their attorneys, to resolve the dispute among themselves in the absence of any third party.

To aid in this negotiation process, where the differences are of a technical nature, the resolution of the technical differences may initially be discussed between the technical personnel of the disputants of all parties, convening to agree on a common technical basis for the solution of the dispute. As an alternative to the in-house engineers and other technical personnel participating in the discussions, each party to the dispute can retain forensic experts to conduct such discussions among themselves and arrive at a mutually agreeable resolution of any technical problems. This writer has recently had the experience of such a meeting whereby forensic engineers, retained by the owner, architect, contractor, and subcontractor met for a half-day and agreed on a significant number of the parameters to the technical aspects of the dispute and then conveyed the results of their meeting to their clients.

The prime purpose of such an internal negotiation is for the parties to settle their differences prior to the establishment of an adversarial position. Such an adversarial position could eventually lead to decisions reached in court or in arbitration. However, if internal negotiations are not successful, the next step would be for the parties to agree to formal mediation of the dispute.

13.3   MEDIATION

To again quote from Poirot in the ASCE Journal:

The preceding process (negotiation) may be considered an internal mediation process. Often times this does not work because the parties involved are too fixed in their positions. When this occurs, external mediation (i.e., a structured negotiation) may provide the solution. The design/construction industry has numerous highly qualified individuals who serve as mediators in construction disputes. There are formal organizations that provide qualified candidates to mediate construction disputes . . . . Mediation is not binding but often provides a forum for both parties to review their own position and reach a settlement. This is by far less costly and more expeditious than the remaining alternatives which are arbitration and litigation. [Poirot 1987]

In her paper, Nelson states:

Mediation is the preferred mechanism because it involves a process that speaks to the business community at large with considerable respect and integrity. Mediation is preferred because it works. Insurance companies which have incorporated ADR and particularly mediation in their risk and litigation management programs have realized some astonishing results. Reasonable business people disagree, and yet, given the opportunity and proper forum, they can come to agreements. Mediation is the ADR mechanism chosen above others when ADR is proposed in a dispute situation. [Nelson, 1987]

Basically, in the mediation process, each party must agree to nonbinding mediation, and a mutually acceptable mediator is selected who will utilize all of his or her influence to enable the parties to agree to a settlement of their differences without attorneys present. The process attempts to avoid formal litigation in the courts or resolution by binding arbitration.

The mediator examines the positions of all parties to the dispute, confers with each of them individually, ascertains points where each party may "give" in their initial positions, and then brings the parties together to arrive at a mutually acceptable compromise. At times, the information furnished the mediator is of a confidential nature and the mediator is then bound not to reveal such matters to the other parties or when all parties are present. Nelson has stated:

> The key to a successful mediation is a sophisticated mediator, not one who is necessarily schooled in the profession of the problem, but one who has knowledge of the dynamics and pressures in a collective negotiation process containing as many or more interests as the parties. The mediator should be someone who is a good conflict resolution manager and sounding board, can prevent distortion of the facts, clarify the parties' respective interests, and enable them to make the appropriate choices.
>
> Selecting a mediator with a particular bias or bent does not help the process. The parties share the cost of the mediator (although the carrier initiating the process will sometimes pay for the initial services required until agreement to the process is secured from all parties). This cost-sharing agreement reinforces the commitment to the process and ensures that none of the parties can "buy" themselves an advocate [Nelson 1987]

The following is cited in Nelson's paper:

> Mediation is and must be voluntary. The mediator's power rests on continuing consent, not authority. The actual participation of the parties whose interests must be satisfied makes the process a powerful way to clear misperceptions, reduce rancor, and provoke a realistic appraisal of the situation. [Phillips and Piazza 1984]

In this writer's opinion, the ideal mediator is one who has technical knowledge of the elements in the dispute and, as Nelson points out, that individual should be a good conflict resolution manager. Typically, such a mediator could be a forensic engineer who has been trained in the mediation process.

Attorneys should not be present as they can be a hindrance, rather than a help, to the resolution process. In this regard, Poirot (1987) states: "By training, background and experience, lawyers are advocates who represent one side or the other. ADR requires a certain amount of dispassionate involvement and a willingness to compromise." Poirot further writes that lawyers can play many important roles in the ADR process, but this writer believes that this should be only in an advisory capacity. Attorneys can be utilized in such roles as drafting the settlement documents after the principals, themselves, have agreed to a resolution of the problem.

Mediation in the construction industry, while used only within the past several years, is successful. Nelson writes that the success factor for cases proposed for mediation is currently at 90%, even without contractual agreement, and that this rate is on the rise. This writer has received a letter from a senior representative of a professional liability insurer who has stated that in one locality, a mediation firm has brought 80% of the cases they assign to the discussion point, and that after the parties are at the table, 72% of those cases settle.

In a further citation from Nelson, "Mediation has been described by one provider as a 'turbocharged' negotiation" (Phillips and Piazza 1984). At times, parties prefer not to utilize the mediation process, usually on advice of their attorneys. Poirot lists the following reasons why formal mediation may not be accepted:

> The financial loss by implementing the solution is too great and one party or the other perceives that a less costly solution will result from legal action.
>
> One party or the other is not willing to admit fault.
>
> The solution is beyond the financial capability of the responsible party and legal action through the courts with possible bankruptcy is the only final settlement process.
>
> Damage to egos and ingrained principles will not allow acceptance of the mediation process. [Poirot 1987]

In spite of these reasons, it is this writer's opinion that a major portion of construction industry disputes can be successfully solved through mediation, a structured negotiation.

13.4 OTHER FORMS OF ADR

There are several other forms of ADR procedures that will not be explained here in detail. These have been discussed by Poirot (1987) and in ASFE's *Institute of Professional Practice*, which is currently in preparation. These include minitrials, dispute resolution boards, and "rent-a-judge." All of these procedures have the intent of reducing time and cost in the resolution of disputes.

One interesting illustration of the use of ADR has been proposed by Donald E. Shepardson, a consulting engineer in California, on behalf of ASFE. This is termed *Third Party ADR by Covenant* (Shepardson 1987). In this concept, a covenant runs with the deed when property is conveyed to a third party. Such a covenant is similar to a right-of-way that is common in many property deeds, and requires that if a claim is made by a new owner (third party) every attempt will be made to resolve this claim by a selected ADR procedure prior to litigation. This concept has been approved by ASFE and by the California Geotechnical Engineers' Association and is currently being used on a trial basis in California.

Finally, mention will be made of a specific ADR procedure that has had widespread interest and approval of several engineering, contractor and legal organizations. This is the "100-day document" concept.

13.5 THE "100-DAY" DOCUMENT

In 1983 the Deep Foundations Institute (DFI), an organization comprising several engineering, general contractor, subcontractor, and materials supplier organizations, recognized the need to avoid the time-consuming and extremely costly conventional litigation process. From DFI emerged the Deep Foundation Construction Industry Roundtable (DFCIR) to develop a document utilizing ADR procedures that could be incorporated into contracts between the owner and the design professional, the owner and the general contractor, and the general contractor and its subcontractors and material suppliers. DFCIR, composed of DFI representatives in addition to representatives of the legal, insurance, surety, and banking industries, met on numerous occasions over a 4-year period to develop what is now known as the 100-day document (DFCIR, 1987).

The "100-day" concept is to have a provision in the respective contract documents for "mandatory discussion, mediation and arbitration of construction disputes."

The document provides for the resolution of a dispute triggered by an event (Day 1) by mandating that the aggrieved party immediately notify the party with whom the aggrieved party has the contract of the nature of the dispute. Mandatory negotiations must take place within a seven-calendar-day period. These negotiations (discussions) are similar to those described in Section 13.2.

In many cases these internal negotiations can resolve the dispute. In the instance where the negotiations are unsuccessful, a mediator, initially named in the contract document, is called in on Day 8 to conduct a nonbinding structured mediation similar to that described in Section 13.3. The mediator has 30 days from the occurrence of the event to conclude a successful mediation, with the parties to the dispute executing a mutually acceptable settlement agreement.

Based on historical experience as mentioned in Section 13.3, 70 to 90% of the disputes can be resolved within this 30-day period. However, if the parties to the dispute, with the aid of a mediator, cannot agree among themselves to resolve the dispute, the services of the mediator are terminated. Immediately one or more arbitrators, also initially named in the contract, are notified to commence binding arbitration in accordance with the rules delineated in the contract. The arbitrators have from Day 30 to

Day 50 (from the inception of the event) to establish a time schedule for the initiation of arbitration proceedings. The binding arbitration will commence from Day 50 and is to be concluded by Day 80. The arbitrator(s) has 20 days in which to report findings and render a decision. The process terminates within 100 calendar days from the commencement of the event that gave rise to the dispute.

The procedure just described has the advantage of avoiding the long and costly litigation in the courts and, if the arbitration provision is invoked, it will be so structured to conclude in no more than 30 calendar days. Without question, the dispute in many instances can be concluded within the negotiation and mediation stages. In no event can the process consume more than 100 calendar days from the date that the dispute arose. The 100-day document is currently being considered on a trial basis in several forthcoming contracts.

13.6 CONCLUSION

The long and costly conventional litigation process is discussed in Section 13.1. It is this writer's opinion that alternate dispute resolution techniques should be employed, where feasible, in a timely manner to properly resolve the issues at hand. These ADR techniques can result in an expeditious settlement, to the satisfaction of all parties to the dispute.

References

American Arbitration Assocation (AAA). 1987. Minutes of Annual Meeting of National Construction Industry Arbitration Committee, AAA, New York, October.

American Society of Civil Engineers (ASCE) Committee for Manual of Professional Practice. *Manual of Professional Practice for Quality in the Constructed Project.* New York: ASCE, in preparation.

Association of Soil and Foundation Engineers (ASFE). *Institute of Professional Practice.* Silver Spring, MD: ASFE, In preparation.

Deep Foundation Construction Industry Roundtable. 1987. Mandatory discussion, mediation and arbitration of construction disputes. Paper presented at Deep Foundations Institute, Sparta, NJ, April.

Lyons, J. 1985. Arbitration: The slower, more expensive alternative? *American Lawyer* (AM. Law Publishing Corp., New York) 7 (January–February):107.

Nelson, S.L. 1987. ADR—A different ritual: An insurer's perspective. *Journal of Performance of Constructed Facilities* (ASCE) 1, No. 4 (November).

Pavalon, E. I. 1987. ADR—Trial lawyer's perspective. *Journal of Performance of Constructed Facilities* (ASCE) 1, No. 4 (November).

Phillips, B., and A. Piazza. 1984. How to use mediation. *Litigation* 10 (Spring).

Phillips, B., and A. Piazza. 1985. Mediation: Putting disputes into a manageable framework. *CPCU Journal* (Chartered Property and Casualty Underwriters).

Poirot, J. W. 1987. Alternative dispute resolution techniques: Design professional's perspective. *Journal of Performance of Constructed Facilities* (ASCE) 1, No. 4 (November).

Shepardson, D. E. 1987. *Third Party ADR by Covenant.* Yorba Linda, CA: ASFE and Californai Geotechnical Engineers Association.

Ward, J. S. 1986. Realities of resolving conflicts during construction. Paper presented at the International Conference on Deep Foundations, Beijing, People's Republic of China, September.

14. Conclusion

KENNETH L. CARPER

The content of this book has been derived from many decades of experience by the contributing forensic experts. The contributors represent a wide variety of specialized disciplines, each using specific investigative procedures and analysis techniques.

Despite the unique characteristics of particular disciplines, all forensic engineers share the common goal of searching for the truth. Joel Hicks (Chapter 6) has written that the forensic engineer must remain flexible to "find the one overriding target, the truth." Lindley Manning (Chapter 5) also states that the function of the forensic engineer is "to search for the truth, and to present that truth in a manner that can be understood by laypersons." Several contributors have emphasized the need for impartiality while conducting an investigation, so that conclusions can be based on the facts.

The forensic engineer deals with the engineering aspects of legal problems. A properly conducted forensic investigation can bring engineering reason into the dispute resolution process, thus contributing to more equitable settlements.

The discipline of forensic engineering has become more visible in recent years. Organizations such as the National Academy of Forensic Engineers have been established, bringing together forensic specialists from many disciplines to work for common goals. The American Society of Civil Engineers has formed the Technical Council on Forensic Engineering. Other professional societies have likewise recognized the important contributions of forensic engineers.

In the past, forensic engineering has been associated almost exclusively with connotations of litigation activity (i.e., the expert witness in the courtroom). While forensic engineers continue to play an important role in litigation, they are becoming increasingly involved in a wide variety of activities that may have little to do with litigation. In civil engineering, for example, the skills and knowledge of the forensic engineer are useful in assessing the performance of the deteriorating infrastructure, and in making recommendations for repair procedures. The civil engineering forensic consultant, with a knowledge of construction practices, materials, and design standards used in past eras, is often called on to assist in the adaptive reuse and historic conservation of significant architectural structures. Forensic engineers have also expanded their role in contributing to the dissemination of information resulting from failure investigation, so that practices and products may be improved.

The practice of forensic engineering is a challenging and rewarding profession. Forensic engineering requires technical expertise to find the facts, communication skills to explain those facts, and professional diplomacy to assist in the resolution of related disputes.

The successful forensic engineer serves society by contributing to an understanding of the causes of failures and accidents. The efforts of the forensic engineer advance the art and science of engineering as the result of learning from failures.

APPENDIX

ICED/ASFE

Recommended Practices

Note: The following document is published with permission from the Association of Soil and Foundation Engineers (ASFE): the Association of Engineering Firms Practicing in the Geosciences, Silver Spring, Maryland. This document has been endorsed by a number of professional societies through the Interprofessional Council on Environmental Design (ICED).

RECOMMENDED PRACTICES FOR DESIGN
PROFESSIONALS ENGAGED AS EXPERTS IN THE
RESOLUTION OF CONSTRUCTION INDUSTRY
DISPUTES[1]

Preamble

Experts are vitally important to contemporary American jurisprudence. They review and evaluate complex technical issues and explain their findings and opinions to lay triers of fact for the latter's consideration in reaching a verdict.

Experts retained by opposing parties may disagree. In all instances, such disagreements should emanate only from differences in professional judgment.

---

[1] For the purposes of this document, the construction industry includes organizations and individuals involved in the design, construction, ownership or use of buildings, public works, and land. The design professionals include architects, engineers, landscape architects, geologists, surveyors, planners, and others.

These recommendations have been developed from the belief that adherence to them will help experts provide to triers of fact substantiated professional opinions unbiased by the adversarial nature of most dispute resolution proceedings. The organizations which endorse these recommendations do not require any individual to follow them.

Recommendations

It is the obligation of an expert to perform in a professional manner and serve without bias. Toward these ends:

1. *The expert should avoid conflicts of interest and the appearance of conflicts of interest*

Regardless of the expert's objectivity, the expert's opinion may be discounted if it is found that the expert has or had a relationship with another party which consciously or even subconsciously could have biased the expert's services or opinions. To avoid this situation, experts should identify the organizations and individuals involved in the matter at issue, and determine if they or any of their associates have or ever had a relationship with any of the organizations or individuals involved. Experts should reveal any such relationships to their clients and/or clients' attorneys to permit them to determine whether or not the relationships could be construed as creating or giving the appearance of creating conflicts of interest.

2. *The expert should undertake an engagement only when qualified to do so, and should rely upon other qualified parties for assistance in matters which are beyond the expert's area of expertise*

Experts should know their limitations and should report their need for qualified assistance when the matters at issue call for expertise or experience they do not possess. In such instances, it is appropriate for experts to identify others who possess the required expertise, and to work with them. Should an expert be asked to exceed his or her limitations and thereafter be denied access to other professionals, and should the expert be requested to continue association with the case, the expert should establish which matters he or she will and will not pursue; failing that, the expert should terminate the engagement.

3. *The expert should consider other practitioners' opinions relative to the principles associated with the matter at issue*

In forming their opinions, experts should consider relevant literature in the field and the opinions of other professionals when such are available. Experts who disagree with the opinion of other professionals should be prepared to explain to the trier of fact the differences which exist and why a particular opinion should prevail.

4. *The expert should obtain available information relative to the events in question in order to minimize reliance on assumptions, and should be prepared to explain any assumptions to the trier of fact*

The expert should review those documents, such as tenders and agreements, which identify the services in question and any restrictions or limitations which may have applied. Other significant information may include codes, standards and regulations affecting the matters in dispute, and information obtained through discovery procedures. If pertinent to the assignment, the expert should also visit the site of the event involved and consider information obtained from witnesses. Whenever an expert relies on assumptions, each assumption should be identified and evaluated. When an assumption is selected to the exclusion of others, the expert should be able to explain the basis for the selection.

5. *The expert should evaluate reasonable explanations of causes and effects*

As necessary, experts should study and evaluate different explanations of causes and effects. Experts should not limit their inquiry for the purpose of proving the contentions advanced by those who have retained them.

6. *The expert should strive to assure the integrity of tests and investigations conducted as part of the expert's services*

Experts should conduct tests and investigations personally, or should direct their performance through qualified individuals who should be capable of serving as expert or factual witnesses with regard to the work they performed.

7. *The expert witness should testify about professional standards of care[2] only with knowledge of those standards which prevailed at the time in question, based upon reasonable inquiry*

When a design professional is accused of negligence, the trier of fact must determine whether or not the professional breached the applicable standard of care. A determination of the standard of care prevailing at the time in question may be made through investigation, such as the review of reports,

---

[2] Standard of care is commonly defined as that level of skill and competence ordinarily and contemporaneously demonstrated by professionals of the same discipline practicing in the same locale and faced with the same or similar facts and circumstances.

records, or opinions of other professionals performing the same or similar services at the time in question. Expert witnesses should identify standards of care independent of their own preferences, and should not apply present standards to past events.

8. *The expert witness should use only those illustrative devices or presentations which simplify or clarify an issue*

The attorney who will call the expert as a witness will want to review and approve illustrative devices or presentations before they are offered during testimony. All illustrative devices or presentations developed by or for an expert should demonstrate relevant principles without bias.

9. *The expert should maintain custody and control over whatever materials are entrusted to the expert's care*

The preservation of evidence and the documentation of its custody and care must be necessary for its admissibility in dispute resolution proceedings. Appropriate precautions may in some cases include provision of environmentally controlled storage.

10. *The expert should respect confidentiality about an assignment*

All matters discussed by and between experts, their clients, and/or clients' attorneys should be regarded as privileged and confidential. The contents of such discussions should not be disclosed voluntarily by an expert to any other party, except with the consent of the party who retained the expert.

11. *The expert should refuse or terminate involvement in an engagement when fee is used in an attempt to compromise the expert's judgment*

Experts are employed to clarify technical issues with objectivity and integrity. Experts should either refuse or terminate service when they know or have reason to believe they will be rewarded for compromising their objectivity or integrity.

12. *The expert should refuse or terminate involvement in an engagement when the expert is not permitted to perform the investigation which the expert believes is necessary to render an opinion with a reasonable degree of certainty*

It is the responsibility of experts to inform their clients and/or their clients' attorneys about the scope and nature of the investigation required to reach opinions with a reasonable degree of certainty, and the effect which any time, budgetary or other limitations may have. Experts should not accept or continue an engagement if limitations will prevent them from testifying with a reasonable degree of certainty.

13. *The expert witness should strive to maintain a professional demeanor and be dispassionate at all times*

Particularly when rendering testimony or during cross-examination, expert witnesses should refrain from conducting themselves as though their service is a contest between themselves and some other party.

Index